Air Handling System Design

Other HVAC Books of Interest from McGraw-Hill

Beaty
Sourcebook of HVAC Details

Caristi
Practical Air Conditioning Equipment Repair

Grimm and Rosaler
Handbook of HVAC Design

Haines
HVAC Systems Design Handbook

Haines and Wilson
Roger Haines on HVAC Controls

Hartman
Direct Digital Controls for HVAC Systems

Levenhagen and Spethmann
HVAC Controls and Systems

Parmley
HVAC Field Manual

Wang
Handbook of Refrigeration and Air Conditioning

Air Handling System Design

Tseng-Yao Sun
Hayakawa Associates
Los Angeles, California

McGraw-Hill, Inc.
New York San Francisco Washington, D.C. Auckland Bogotá
Caracas Lisbon London Madrid Mexico City Milan
Montreal New Delhi San Juan Singapore
Sydney Tokyo Toronto

Library of Congress Cataloging-in-Publication Data

Sun, Tseng-Yao.
Air handling system design / Tseng-Yao Sun.
p. cm.
Includes index.
ISBN 0-07-062571-9
1. Heating—Equipment and supplies. 2. Ventilation—Equipment and supplies. 3. Air conditioning—Equipment and supplies. I. Title.
TH7345.S83 1994
697—dc20 93-27079
CIP

1 2 3 4 5 6 7 8 9 0 DOC/DOC 9 9 8 7 6 5 4 3

ISBN 0-07-062571-9

The sponsoring editor for this book was Robert W. Hauserman, the editing supervisor was David E. Fogarty, and the production supervisor was Pamela A. Pelton. This book was set in Palatino by McGraw-Hill's Professional Book Group composition unit.

Printed and bound by R. R. Donnelley & Sons Company.

This book is printed on recycled, acid-free paper containing a minimum of 50% recycled de-inked fiber.

Contents

Part 4. Other System Considerations

Preface

In the past I have written many articles and papers on specific HVAC subjects to share my thoughts and experiences with my colleagues. For the past fifteen years, however, I have had this burning desire to write a book on the various types of air handling systems. Every time I got a little more serious, the excuses for not writing the book always outweighed the reason for writing it. Thanks to Mr. Robert Hauserman of McGraw-Hill who gave me the opportunity and the much-needed encouragement, this book has finally become a reality.

This book discusses the practical applications in selection and design of various air handling systems. The text is divided into four parts. Part 1 deals with basic and general concerns of air handling systems. Part 2 discusses a variety of constant volume air handling systems. Part 3 goes into a variety of variable volume and hybrid air handling systems. Part 4 deals with a few unique system applications which relate air handling systems with other concerns in the air-conditioning design.

I have often thought that the material discussed in this book is one of the missing links between the academic community and the practicing professionals in the HVAC field. Students who learned heat transfer, thermodynamics, and air-conditioning theories from the university often lacked a general understanding of what goes on in the practicing field. It is my hope that the material introduced in this book can be used to develop a HVAC curriculum so that college graduates who enter the HVAC field will be better acquainted with the practicing end of the HVAC field.

I would like to express my sincere gratitude to my friends and colleagues: Tom Romine, who for a long time encouraged me to write this book and provided much guidance and editorial comments to the text; Ray Alvine, who read the manuscript and gave me many suggestions, criticisms, and guidance; and Lynn Bellenger, who gave me the much-needed help and encouragement during the early planning stage of this major endeavor.

My thanks also go to Kathy Thornton of our office, who for years has helped me improve my English and provided substantial help in putting together the manuscript of this book. Thanks to Bob Okajima of our office, who helped put all the figures used in this book on CADD. Special thanks to my son, Larry, who is also my friendly competitor, for his many editorial comments and suggestions.

Tseng-Yao Sun

Air Handling System Design

PART 1
General

1

Introduction

This book presents an inventory of various types of central air handling systems. The air handling system is an important link in a comfort air-conditioning system. The system takes the heating and cooling effect generated from such primary sources as chillers and boilers and provides heating and cooling directly to the conditioned spaces. The system also must be designed to satisfy the important ventilation and filtration requirements for proper indoor air quality control in the conditioned spaces. For special occupancies requiring humidity control, the selection of the air handling system also must take the humidity requirement into consideration.

This book concentrates on the heating and cooling aspects of the different air handling systems. Humidity control, filtration requirements, and their influence on the selection of air handling systems are also discussed. It is not the intent of this book to recommend one system over another. To the contrary, it is believed that every system included in this book has its own merits and applications. Some systems may be more popular than others, while other, seldom used systems may have their own unique applications where popular systems are not satisfactory. The goal of this book is to discuss comprehensively all commonly used air handling systems, as well as some that are not so commonly used, in one source so that their merits and drawbacks can be compared and evaluated easily.

A few of the air handling systems discussed in this book may be unique and may even appear impractical today to some heating, ventilation, and air-conditioning (HVAC) engineers. These systems, however, are included here for their specific applications, for historical value, and more important, for stimulating new system development ideas.

Many different types of air handling systems have been developed in the past. Undoubtedly, new air handling systems can still be developed for unique applications. Some air handling systems may have better thermal and/or mechanical efficiency than others. As a rule, however, central air handling systems serving multiple occupancies will always have thermal and mechanical inefficiencies. The aim of the "game" is to minimize system inefficiency, achieve design comfort, and not jeopardize the long-term health and safety of building occupants. In many instances, multiple types of air handling systems are used in large building complexes or even in one large building to satisfy different functional needs and to minimize the energy waste due to different system inefficiencies.

In discussing air handling system selection, system efficiency and operational considerations are often gauged against installation costs. Systems that have lower installation costs but are less efficient or require more maintenance are often selected in an effort to "bring the project within budget." During the developmental stage of a project, the importance of installation costs almost always outweighs system efficiency and operational considerations. Many times, the installation costs are not limited to the air-conditioning system only. Increasing ceiling height or reducing the floor-to-floor height at the expense of reducing duct size is a common occurrence. A classic example is the use of a single-duct reheat system with high-velocity, high-pressure duct distribution to save sheet metal cost as well as building height. The penalty in this case, of course, is the high thermal energy waste and high-horsepower requirements for the system to operate under high-pressure conditions.

The proper prospective one must keep in mind when evaluating system design and selection is that installation costs are only a one-time expense. On the other hand, the added thermal and transportation energy costs of an inefficient air-conditioning system are ongoing expenses that are always escalating as energy resources are depleted and the cost of energy rises.

In the 1980s, many state and federal government agencies developed rules and regulations in an effort to curb this shortsightedness. More energy-efficient systems have been used more often, and the general public is now aware that energy is a depletable commodity. A better understanding of system efficiency, or inefficiency, among commonly used central air handling systems becomes an essential prerequisite for an HVAC engineer. It is hoped that in the future, with everyone clearly understanding where and how energy is wasted in HVAC design, our industry will be made self-regulating so that we will not be governed by complex rules and regulations which, by nature, cannot be all-inclusive and sometimes create confusion and waste of "human energy."

2
The Basics

Relationship Between Load and Supply Air

In designing an air-conditioning system, one must first calculate the required heating and cooling loads for each conditioned space. The load calculation procedure, which will not be discussed in this book, can be found in the *ASHRAE Handbook of Fundamentals,* as well as in various air-conditioning textbooks.

The basic relationship between the space load and the conditioned air required to satisfy the space load can be expressed as

$$Q_s = 1.1 \times \text{CFM} \times DT \tag{2.1}$$

where Q_s, the sensible load of a conditioned space at a given time (in Btu/h), must be met by a certain amount of conditioned air, CFM (cubic feet per minute) at a temperature DT degrees above, for heating, or below, for cooling, the space temperature; 1.1 is the product of the specific weight of dry air at sea level, 0.075 lb/ft^3, the specific heat of moist air at 55°F dry bulb, 24.4 Btu/(°F·lb dry air), and 60 min/h.

Rearranging Eq. (2.1), the supply air quantity CFM required to satisfy the peak space sensible load $Q_{s,\max}$ can be calculated using Eq. (2.2):

$$\text{CFM} = \frac{Q_{s,\max}}{1.1 \times DT} \tag{2.2}$$

where DT is a preassigned temperature differential between the room temperature to be maintained and the temperature of the supply air.

Equation (2.1) indicates that the heating or cooling load of a conditioned space is a function of the supply air quantity and the temperature difference between the supply air and the space condition. From this one may conclude that air handling systems can be categorized into two major groups:

1. One group of systems will keep the supply air volume (CFM) constant and relies on varying the temperature difference *DT* to meet the varying load requirements in the conditioned spaces.
2. The second group of systems will keep the temperature difference *DT* constant and vary the supply air quantity (CFM) to satisfy the load variation.

These two major groups of air handling systems are commonly referred to as *constant air volume systems* and *variable air volume systems*, respectively. Equations (2.1) and (2.2) and the terms *Q*, CFM, and *DT* will be referenced throughout this book.

Difference Between Heating Energy and Cooling Energy

The terms *heating energy* and *cooling energy* will be used often in this text. It is necessary to know the difference between these two terms. Heating is an energy-replacement process. Heat energy is generated by burning fuel or using electricity, and the energy thus generated is supplied to conditioned spaces via a transporting medium, e.g., steam, hot water, and/or hot air, to replace the heat loss from the controlled environment. Cooling, on the other hand, can be viewed as a pumping process.

In the cooling process, heat is removed from a conditioned space at a lower temperature level and rejected to a higher-temperature environment outside. Refrigerant is used as the transporting medium, and the refrigerant absorbs the heat in the air from the conditioned space directly or indirectly (via chilled water) and rejects it to the outside using power to the refrigerant compressor or heat to the absorption chiller as the prime mover of the pumping action.

The difference in energy use between heating and cooling processes can be illustrated in another way by using the following example. For a conditioned space having a heating load of 12,000 Btu/h, an equal amount of heat, 12,000 Btu/h, has to be generated and supplied to the conditioned space. On the other hand, for a conditioned space with 12,000 Btu/h cooling load, 1 ton of refrigeration is needed to remove the heat. Using an electric-drive refrigerant compressor, which uses approximately 1 kW/ton,* the equivalent heat value is 3146 Btu/h, considerably different from the 12,000 Btu/h required for the heating process.

Single-Zone Air-Conditioning

For an air handling system serving a single air-conditioned space, often referred to as a *single-zone system,* the cooling and heating requirements can be readily achieved by installing the required heating and cooling sources in the air handling unit to perform the functional needs with a blower moving conditioned air to and from the space (Fig. 2.1). When the space requires heating, the heating source, generally a heating coil, a gas furnace, or an electric-resistance heater, will supply heat to the conditioned airstream and raise its temperature to satisfy the space-heating load. When the space requires cooling, the cooling source, generally a chilled-water or refrigerant (direct-expansion, or DX) cooling coil, will cool the conditioned airstream and lower its temperature to satisfy the space-cooling load. Since the space load varies as a function of time, varying amounts of heating or cooling energy are needed to match the heating or cooling loads. When the space needs neither heating nor cooling, a *no-load condition,* or—more often the case—when heat loss and heat gain to the conditioned space neutralize each other, neither the heating source nor the cooling source will be activated. The blower can stop under no-load conditions or continue to run to provide needed ventilation.

Without specific involvement of humidity control, which will be discussed in detail in the reheat application, a single-zone system will require either heating or cooling at any given time. There is no need for heating and cooling energy to be available at the same time. All heating or cooling energy generated from the source will be used to match the space load. Thus, in general, a single-zone air-conditioning system does not have any thermal energy wastage.

*The kilowatt per ton value, including auxiliaries such as pumps, cooling towers, or condenser fans, varies from less than 1 kW/ton for centrifugal chillers to as high as 1.5 kW/ton for air-cooled direct-expansion compressors. For absorption chillers, the heat requirement is considerably different.

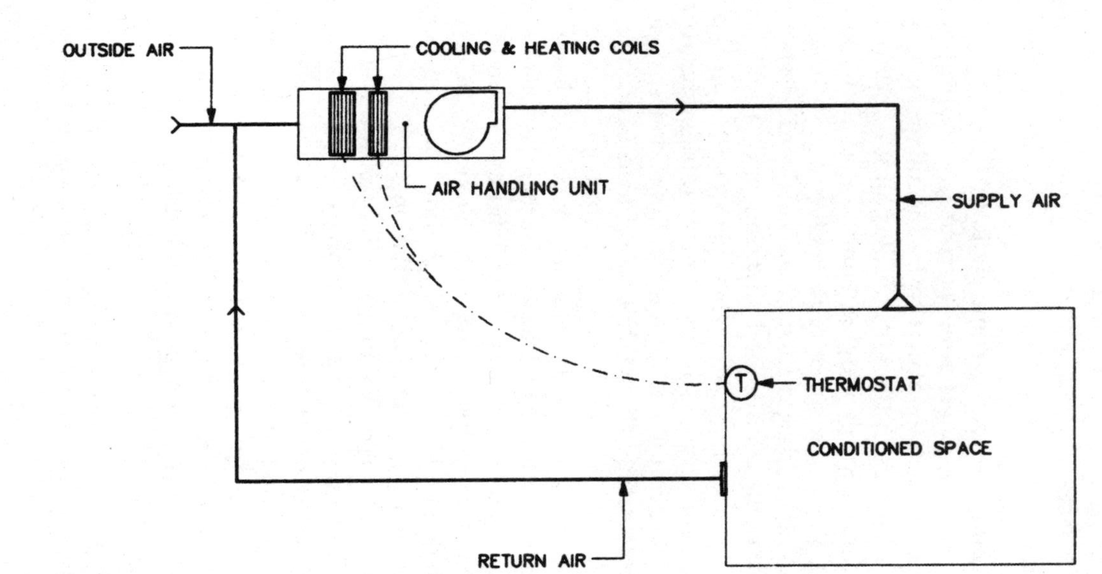

Figure 2.1. Schematic of a single-zone air handling system.

The Need for a Variety of Multizone Air-Conditioning Systems

Unfortunately, a building almost always contains multiple spaces with different occupancy functions and different load profiles, as shown in Figure 2.2. A single-zone system will not be able to satisfy different needs in different spaces at any given time. It is often impractical to install a great number of single-zone systems each serving a conditioned space. For this reason, a variety of central air handling systems with the capability of satisfying different needs in different spaces simultaneously have been developed to serve buildings with complex space requirements.

Different types of central air handling systems are developed for a variety of purposes and reasons. Each type of system may have its own applications and limitations. The type and availability of heating and cooling energy sources, the required mechanical room and ceiling space, the installation costs, maintenance involvement and operating costs of each system are all different. The design engineer must carefully compare and evaluate different types of systems that are suitable for the project before rendering a decision on which system is best suited for the project.

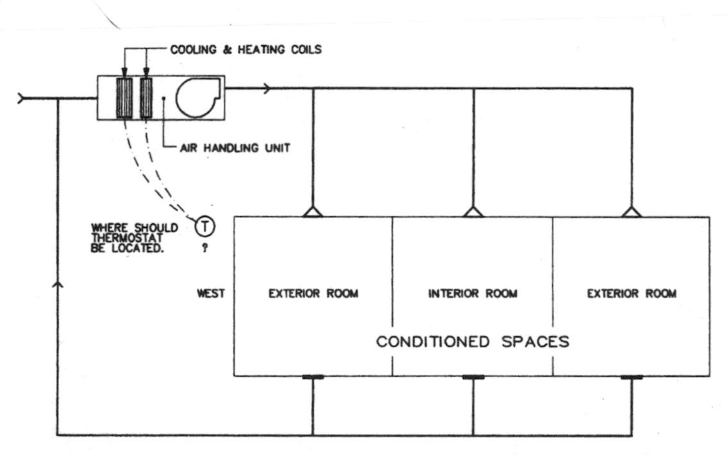

Figure 2.2. Single-zone system serving multiple spaces with different load profiles.

3

Psychrometrics

A clear understanding of the thermodynamic properties of moist air, commonly referred to as *psychrometrics*, is an important prerequisite to selecting and designing an air handling system. Figure 3.1 is an abbreviated psychrometric chart for the normal temperature range of air-conditioning applications and for standard barometric pressure at sea level.

While supply air quantity in an air handling system is directly related to sensible load, see Eq. (2.1), latent load plays an important role in comfort condition in an air-conditioned space. The relationship between the latent load Q_l (in Btu/h) and the supply air quantity CFM to a conditioned space can be expressed as

$$Q_l = \mathrm{CFM} \times 4840 \times DW \tag{3.1}$$

where 4840 is the product of the specific weight of dry air at sea level, 0.075 lb/ft^3, the approximate energy content of 75°F air at 50 percent relative humidity, less the energy content of the water at 50°F, 1076 Btu/lb, and 60 min/h. *DW* is the difference in humidity ratio, in pounds of moisture per pound of dry air, between the supply air and the room condition.

Total load Q_t (in Btu/h) is the sum of sensible load and latent load and can be expressed as

$$Q_t = \mathrm{CFM} \times 4.5 \times DH \tag{3.2}$$

where 4.5 is the product of 0.075 lb/ft^3 and 60 min/h, and *DH* is the enthalpy difference (in Btu/lb dry air) between the supply air and the room condition.

In a normal air-conditioning system, as shown in Figure 2.1, most of the air supplied to the conditioned space is returned to the air handling unit. Outside air is introduced at the air handling unit to satisfy the ventilation requirement. The mixture of returned and outside air is

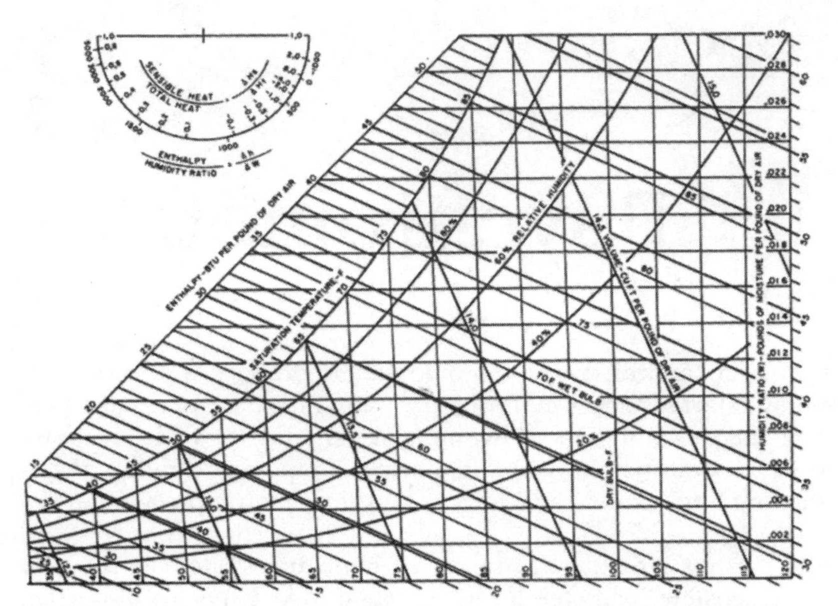

Figure 3.1. Abbreviated psychrometric chart. (*Reprinted with permission from Cooling and Heating Load Calculation Manual, ASHRAE, 1979.*)

then sent through a cooling or heating apparatus, where heat and moisture are removed or added to maintain the desired room condition.

Figure 3.2 illustrates the simplified psychrometric plot of an air-conditioning process corresponding to the system shown in Figure 2.1. Room air, indicated as point *A* on the psychrometric chart, is returned to the air handling unit and mixed with a required amount of outside air, point *B*. The mixed air, point *C*, flows through the cooling coil along the process line *CD*. Conditioned air leaving the cooling coil at point *D* is delivered to the conditioned space and absorbs the space load following process line *DA* to complete the cycle. The slope of line *CD* indicates that in addition to lowering the temperature of the conditioned air, which is due to a sensible heat-removal process, moisture is also removed from the air, which is a latent heat-removal process. The slope of line *DA* represents the effect of room sensible heat ratio (SHR), which is the ratio of room

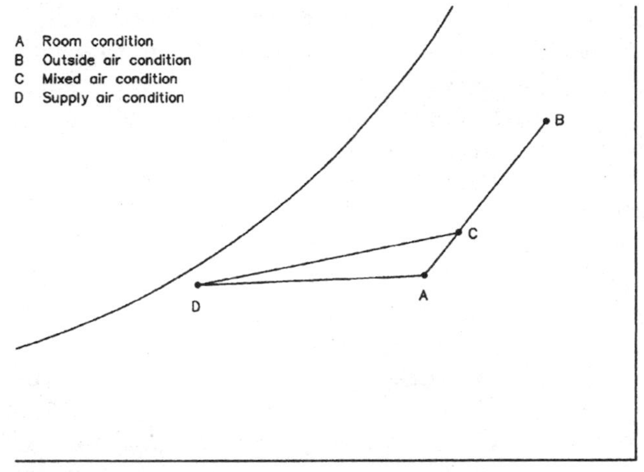

Figure 3.2. Psychrometric process of a simple air-conditioning system.

sensible load to the sum of room sensible and latent loads. Higher latent load in a space, which means higher amounts of moisture released into the conditioned space by the occupancy, will lower the room SHR and cause a steeper slope of the room SHR line (line *DA*). For a conditioned space with no latent load, line *DA* will be a horizontal line.

In addition to using the psychrometric chart, relationships between various psychrometric properties (dry bulb temperature, wet bulb temperature, relative humidity, humidity ratio, enthalpy, and dewpoint), can be obtained by using a psychrometric subroutine computer program.* Using the subroutine, psychrometric analysis can be performed for any air-conditioning process.†

When performing a load calculation, the engineer generally specifies not only the design temperature but also the relative humidity (RH) to be maintained in the conditioned space. A common reference for this design condition is 75°F dry bulb and 50 percent relative humidity. The ability to maintain the design relative humidity in the conditioned

6*"Psychrometric Subroutine Using ASHRAE Algorithms," HPAC, October 1971. See Appendix A.

†"Psychrometric Analysis of AC Systems by Computer," HPAC, November 1971. See Appendix B.

space, however, is a function of the outside air condition, latent load in the space, and performance of the cooling coil. It is important to recognize that 75°F dry bulb and 50 percent relative humidity is a common reference. In arid climates, dry outside air may prevent the system from reaching 50 percent relative humidity in the conditioned space without adding humidity to the air. With the aid of a psychrometric chart, this phenomenon can be clearly illustrated, as shown in Figure 3.3. In plotting an air-conditioning process on the psychrometric chart, the process loop must be closed and in equilibrium. Note that the design condition cannot be maintained because the dry ventilation air will not allow the process loop to close unless the relative humidity in the conditioned space is lowered to allow the process to reach an equilibrium.

The design engineer should recognize that the general goal for the majority of the comfort air-conditioning processes is to control space temperature. The humidity level in an air-conditioned space is determined by the performance of the cooling coil, and affected by weather conditions and the sensible heat ratio of the conditioned space. For a critical space, where the humidity level must be maintained at all times, dehumidification control such as the use of a reheat or chemical dehumidification system, together with the humidifier will be required in order to cope with the changing weather conditions and room sensible heat ratio.

Furthermore, one should be aware that for a central air handling system serving multiple thermostatically controlled spaces, while the return air temperature from each space may be the same (assuming that the thermostats are set at the same temperature), relative humidity levels from these spaces may differ since the sensible heat ratio for each thermostatically controlled space can be different. The type of air handling system chosen, reheat verses mixed air systems, also plays an important role in determining the relative humidity level of the return air. The actual relative humidity of the air returning to the air handling system will be the weighted average of the relative humidity of the return air from all conditioned spaces.

The examples in Figures 3.2 and 3.3 illustrate basic and simplified air-conditioning processes associated with a simple air handling system. Psychrometric plots of various processes associated with the air handling systems discussed in this book will be introduced to give the reader a more complete understanding of the air-conditioning sys-

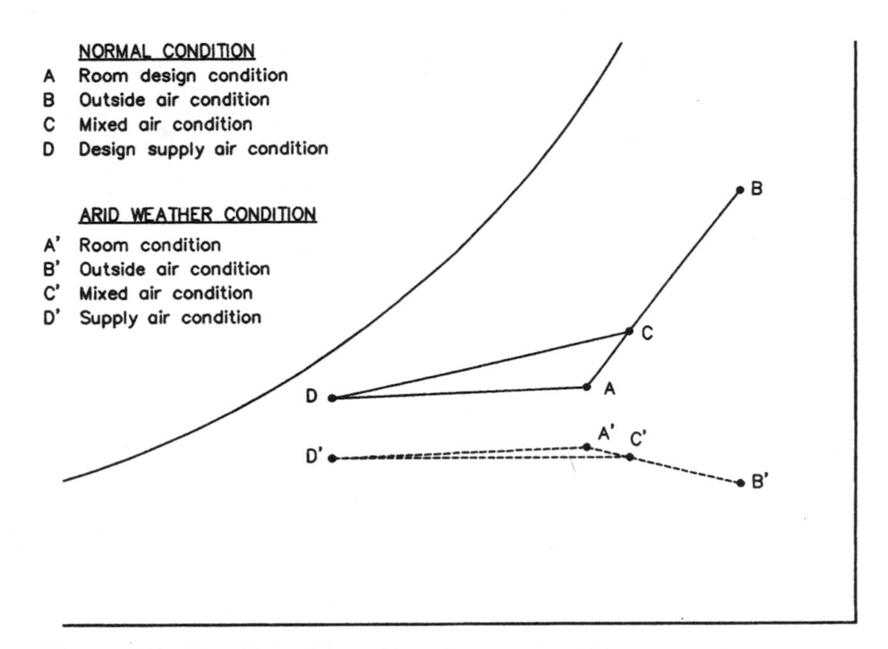

Figure 3.3. The effect of outside air to an air-conditioning system.

tems. The principle of psychrometrics, although not discussed in depth here, can be found in the *ASHRAE Handbook of Fundamentals,* the *ASHRAE Cooling and Heating Load Calculation Manual,* and various air-conditioning textbooks. Readers are advised to familiarize themselves with the psychrometrics in order to visualize a more complete picture of the air-conditioning processes discussed in this book.

4 Heat to Return Air and Fan Heat

Many air-conditioning systems serving commercial buildings do not use a ducted system to transport return air back to the air handling unit. Instead, the ceiling cavity (the space between the top of the ceiling construction and the bottom of the floor or roof structure above) is used as a return air plenum. Grille openings are placed in the finished ceiling to allow the room air to flow into the ceiling cavity and migrate toward the air handling unit. For air handling systems serving multiple floors, a ducted return air riser collects return air from each floor and returns it to the air handling unit. Sometimes, a vertical shaft without a sheet metal duct also can be used to convey the return air.

Recessed light fixtures mounted in finished ceilings give out heat to the ceiling cavity as well as to the room. That portion of lighting heat entering the conditioned space will become a part of the room air-conditioning load. However, with use of a return air plenum, that portion of the lighting heat released into the ceiling cavity will be absorbed by the return air as the air flows over the back of the light fixtures in the ceiling cavity. This portion of the lighting heat does not enter the room and thus cannot be counted as part of the room load. However, it is important to note that lighting heat to the return air will become a part of the air-conditioning system load.

Heat gain through the roof has to pass through the ceiling cavity before becoming part of the room load. In route, a portion of the heat gain through the roof will be picked up by the return air and will never enter the conditioned space below. This is also true for heat gain through that portion of exterior walls which is above the ceiling line. Return air enters the ceiling grille at room temperature. The air temperature will rise as the return air absorbs the heat in the ceiling cavity from lights, roof, and walls on its way toward the air handling unit.

Similarly, for an atrium or a room with an exceedingly high ceiling, because warm air rises, the upper parts of the conditioned space tend to be warmer than the room design temperature. If return air openings in these spaces are located at a higher level, a portion of the heat gain through the atrium skylight or the upper portion of the clerestory will be picked up by the return air at higher temperature before these heat gains become part of the room load.*

Fans used in air-conditioning add heat to the system which will become part of the air-conditioning load. This load causes a temperature rise of the conditioned air and should be accounted for in the cooling-load analysis. The magnitude and appearance of the temperature rise is a function of the location of the fan in the air handling unit (draw-through or blow-through arrangement) and the location of the motor and drive (in or out of the airstream). Heat to return air and fan heat are generally system loads and should not be considered as a part of room loads. These loads are represented on a psychrometric chart as heating processes. Figure 4.1 shows a psychrometric chart of a blow-

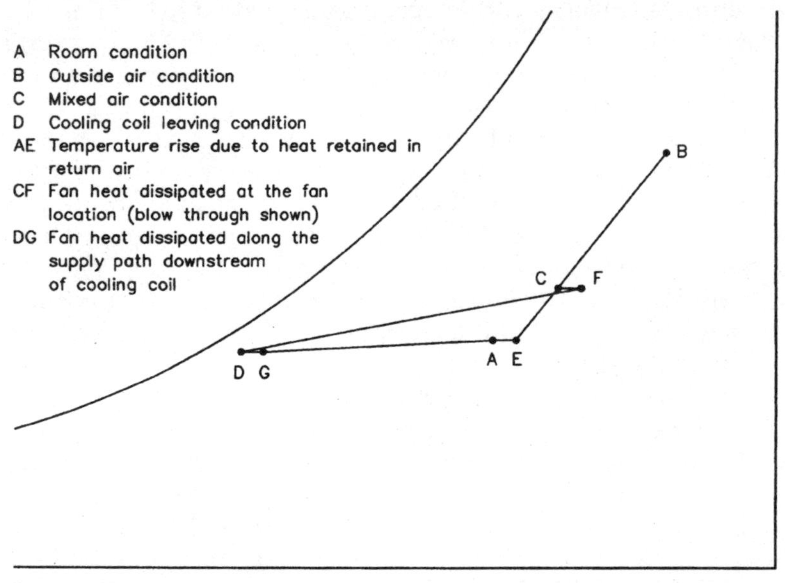

Figure 4.1. Psychrometric plot of an air-conditioning system, including load to return air and fan heat.

*It is a common practice in energy conservation to place the return air openings at a lower level and to allow the temperature at the top of the atrium (above the occupied zone) to rise above the comfort level.

through system, including the heat to return air and fan heat distribution. Note that the temperature rise due to fan heat occurs at two places. A portion of the temperature rise occurs at the fan, which in this case is at the upstream side of the cooling process. The other portion occurs downstream of the cooling coil, which represents the temperature rise caused by the reduction in velocity pressure as the air travels along the supply duct system.

5 Economizer Cycle

An *economizer cycle* is an option in an air handling system design commonly used to save cooling energy. The design scheme takes advantage of the cooler outside air in mild and cold climates to supplement or satisfy the cooling needs.

Interior spaces of a building require cooling year-round. When the outside air temperature drops below the required supply air temperature, the required cooling load can be met be mixing varying amounts of cool outside air with return air. Under this condition, the entire refrigeration plant can be shut down. Even when the outside air temperature is higher than the required supply air temperature but lower than the room temperature, the refrigeration load can be significantly reduced by replacing the higher-temperature return air with cooler outside air, provided that the outside air humidity is not exceedingly high. The use of an economizer cycle to conserve cooling energy is quite practical in colder climates. An economizer cycle can be applied to most central air handling systems, and it is especially effective on single-duct systems. For some mixed air systems, where temperature control relies on mixing two airstreams at different temperatures, application of the economizer cycle may become questionable.

Economizer Cycle Design

Additional fans are generally needed for an economizer cycle. In an air handling system without an economizer, return air is drawn back from the conditioned space to the air handling unit. Negative pressure exists in the mixed air plenum, where return air is mixed with the minimum amount of outside air required for ventilation. The pressure drop in the return air system is borne by the suction of the supply fan. With the economizer cycle, when the system requires 100 percent outside air, a maximum outside air damper will open and a return air damper will close at the mixed air plenum of the air handling unit to allow 100 percent outside air to enter the system. A relief damper at the air handling unit will open to allow the blocked return air to expel to the outside. Under this condition, unless the pressure drop across the return air passage is negligible, which allows relief, the pressure will build up in the conditioned space. For this reason, return or relief fans are generally installed in conjunction with the economizer cycle design.

Two variations of economizer cycle design are shown in Figures 5.1 and 5.2. The economizer design shown in Figure 5.1 uses a return fan to draw air back from the conditioned space and discharge the air into an economizer section. Depending on the outside air condition, the return air can either be returned to the system or relieved to the outside so that outside air can be introduced into the system. This type of economizer cycle design can be applied to both constant volume and variable volume systems. The economizer design shown in Figure 5.2 utilizes a relief fan. In this case, the supply fan bears the duty of drawing return air back to the air handling unit. The relief fan is activated only when the economizer cycle is in operation. During economizer operation, the relief fan extracts varying amounts of return air and discharges it to the outside to allow more outside air to be introduced into the system without overpressurizing the conditioned space. In this case, the relief fan must operate through a varying volume range. To avoid variable volume control, sometimes multiple fans are used to stage the economizer cycle operation. This type of economizer design is generally limited to variable volume system applications. The use of a relief fan may provide more design flexibility, since the relief fan, particularly if multiple relief fans are used, does not have to be centrally located.

For constant volume systems, the economizer controls the temperature leaving the economizer only. For variable volume systems, however, in addition to temperature control, the capacity tracking between the supply and return or relief fans also must be taken into consideration. This is necessary in order to maintain proper pressurization of the conditioned space. More important, throughout the variable volume operating range, the minimum amount of outside air required to satisfy the ventilation requirements must be maintained.

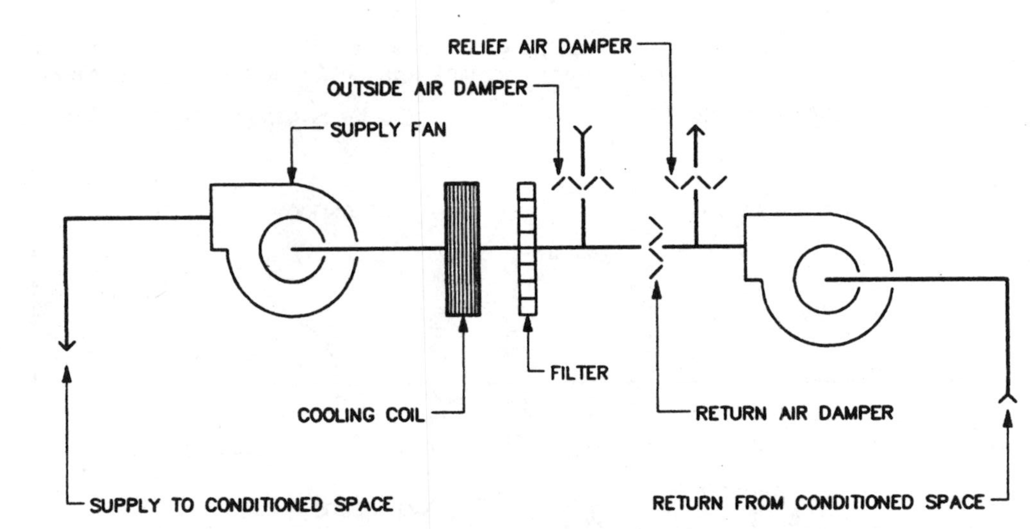

Figure 5.1. Economizer cycle with return fan. The unit shown has draw-through cooling only; other air handling systems are similar.

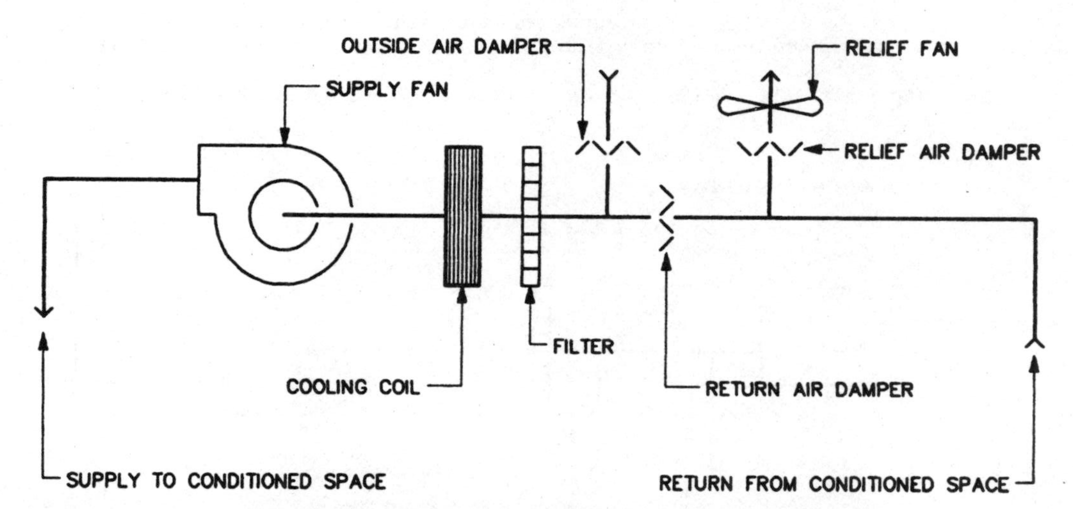

Figure 5.2. Economizer cycle with relief fan. The unit shown has draw-through cooling only; other air handling systems are similar.

Economizer Cycle Control

There are control variations to determine when to use outside air in an economizer cycle. Figure 5.3 shows several psychrometric charts illustrating various ways to control the starting point of using outside air in an economizer. The decision of whether to use an economizer and what type of control to use generally depends on the evaluation of weather patterns, indoor design criteria, and the balance between the additional first cost of the economizer cycle versus potential energy savings.

As pointed out in Figure 5.3, there are unfavorable weather conditions where an economizer cycle should not be used. Generally, the proper way to control an economizer cycle is to use both dry bulb temperature and enthalpy or wet bulb temperature measurements. It

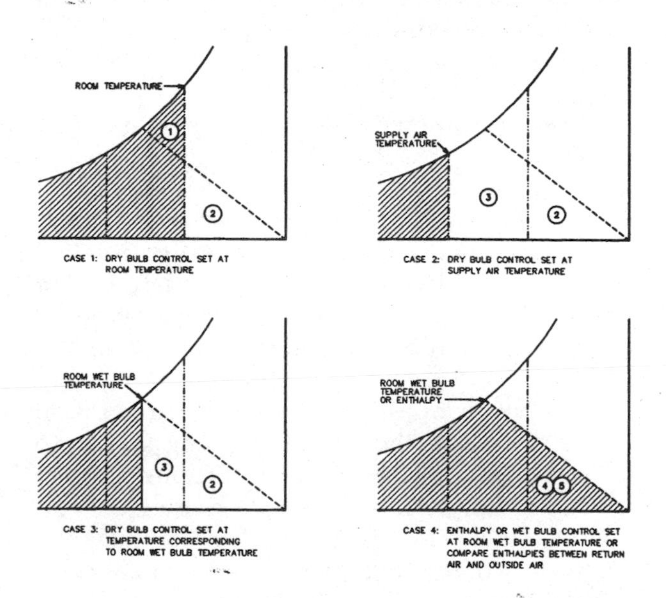

Figure 5.3. Economizer cycle control. Shaded areas indicate economizer cycle operation. *Notes:* (1) Potential energy waste when outside air is in this high humidity range. (2) Potential energy saving not utilized when outside air is in this low-humidity region (see also Notes 4 and 5). (3) Potential energy saving not utilized when outside air is in this region (see also Note 4). (4) Low humidity in this region may have the adverse affect of causing the humidity in the conditioned space to become too low. (5) Sensible cooling of high outside air temperature may use more energy than the sensible cooling of mixed air.

should be emphasized that using either dry bulb temperature or enthalpy alone to control economizers may have adverse effects on energy consumption and humidity control.

Application of the Economizer Cycle

Comprehensive discussions on economizer cycle applications are included in the documentation of most of the air handling systems discussed in this book. One must remember, however, that an economizer operation is a design option, and not all air handling systems are suitable for economizer cycle application. For certain air handling system designs, there will be no advantage to incorporate an economizer; further, in some air handling systems, the use of an economizer cycle may even jeopardize system operation.

Outside air can be extremely dry in the winter time, when use of an economizer cycle is beneficial for temperature control. For spaces requiring humidification, however, the use of an economizer will increase the humidification load. Under this circumstance, the cooling-energy savings from using the economizer cycle should be weighed against the additional cost for humidification. Such an analysis also may prove that other energy-conserving measures, such as a refrigeration cycle heat-recovery scheme, could be used more effectively in lieu of an economizer cycle.

An economizer cycle may not be applicable where a refrigeration cycle heat-recovery scheme is used in the air-conditioning system to conserve energy. The heat-recovery design is an attempt to use rejected heat from the refrigeration cycle to conserve heating energy. The economizer cycle, on the other hand, is designed to minimize the use of refrigeration to conserve cooling energy. The economizer cycle is most effective when the refrigeration plant is shut down, but heat-recovery systems rely on the operation of the refrigeration plant to transfer the rejected heat to satisfy heating needs. Although the two schemes appear incompatible with each other, they can be used harmoniously together to obtain maximum operating efficiency. In most cases, however, it will be difficult to justify the construction costs to install both schemes in an air-conditioning system. See Chapter 32 for more detailed discussion.

6
Single-Zone Systems

The single-zone system is the air handling system in its simplest form (see Fig. 2.1). As stated in Chapters 1 and 2, the system takes the heating and cooling effect generated from the primary sources and provides heating and cooling directly to a conditioned space. By definition, a single-zone system responds to one space temperature controller only, generally a room thermostat. The system will provide varying amounts of cooling or heating to satisfy the setting of the room thermostat. When the room thermostat is satisfied and there is no load in the space, neither the heating source nor the cooling source will be activated, and the air handling system will simply provide ventilation for the space. If the space is unoccupied, most often the system will be shut down to conserve energy.

A single-zone system can serve a single room or a group of rooms with similar load characteristics. In the later case, the thermostat can be located in any one of the rooms or, sometimes, in the common return airstream. It would be a misapplication, or at least a compromise, if a single-zone unit were designed to serve a group of rooms with different load characteristics, as shown in Figure 2.2. It is obvious in this case that there is no place to locate the thermostat and in turn satisfy all rooms at any given time. As shown in Figure 2.2, placing the thermostat in the east-facing room will cause the room facing west to become too warm in the afternoon. On the other hand, placing the thermostat in the west-facing room will overcool the room facing east in the afternoon. Placing the thermostat in the return air may minimize the discomfort but still will not satisfy each individual room condition. For this reason, more complex central air handling systems are developed to have multiple thermostatically controlled spaces served from a single air handling system.

Sometimes, as a compromise, the design engineer or the client may feel that using a single-zone system with the thermostat located in the return airstream or in one of the rooms that is more important than the others is adequate. In this case, however, it is very important for the engineer and client to have a mutual understanding that this is a compromise and to have complete knowledge of the possible consequences.

Control of Single-Zone Systems: Coil Control

There are two methods of controlling the cooling or heating capacity of a single-zone system. The most common method is to control the cooling or heating source to allow the required amount of energy to be supplied to the system. With a single-zone air handling unit using chilled- and hot-water coils, the control valve for the chilled-water coil will modulate open or close as the cooling demand of the system increases or decreases, respectively. And as the cooling season changes into the heating season, the chilled-water control valve will close and the hot-water control valve for the heating coil will modulate open to satisfy the heating load.

Figure 6.1 shows the psychrometric process of a single-zone air handling system with coil control operating under full and partial load conditions. Note that under partial load conditions, since the cooling coil is not operating at its full capacity, the ability of the coil to extract moisture out of the conditioned air is reduced. As the cooling load in the conditioned space is reduced, the cooling coil performance gradually changes from a cooling and dehumidifying process to a sensible cooling process only. If the room sensible heat ratio is low and the outside air relative humidity is high, the humidity in the conditioned space may rise above the design range.

Control of Single-Zone Systems: Face-and-Bypass Control

Another way to control the cooling and heating capacity of a single-zone system is the use of face-and-bypass control, sometimes referred to as *mixed-air control.* In this case, the capacity of the system is con-

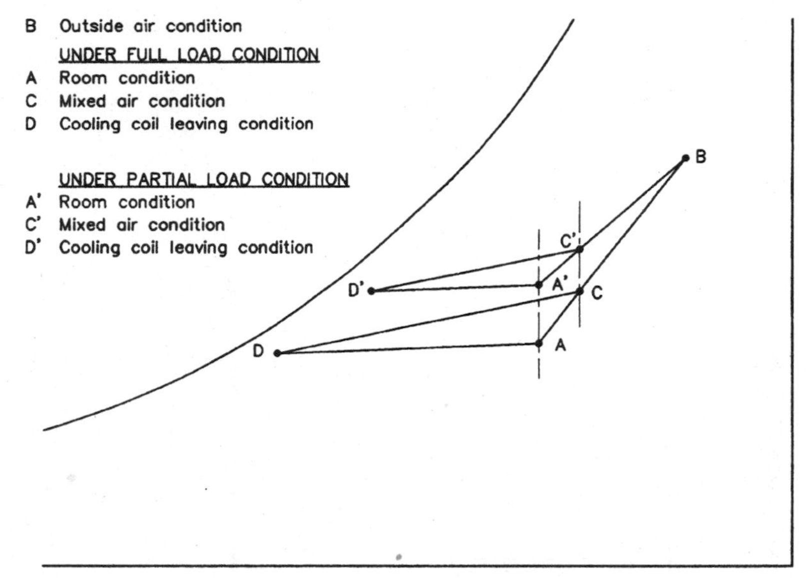

Figure 6.1. Psychrometric plot of a single-zone unit with coil control under partial load conditions.

trolled by a set of dampers, as shown in Figure 6.2. The damper located in series with the cooling and heating coils is called a *face damper*. Another damper located in an air passage parallel to the coils is called a *bypass damper*, and this damper will allow a controlled amount of unconditioned air to pass through.

With this scheme, the modulating control valves used for coil control will be replaced by solenoid valves and used for seasonal changeovers only. In the cooling season, the solenoid valve for the chilled-water coil will be open to allow full chilled-water to flow through the cooling coil and the hot-water valve will be closed. A room thermostat will modulate the positions of the face and bypass dampers to mix proper amounts of airflow through the cooling coil and the bypass path. When the system demands full cooling, the face damper will be fully open and the bypass damper will be closed. As the cooling demand decreases, the thermostat will modulate the face damper to close and the bypass damper to open to satisfy the cooling needs. In the heating season, a similar sequence will apply in the heating cycle with the hot-water valve opened and the chilled-water valve closed.

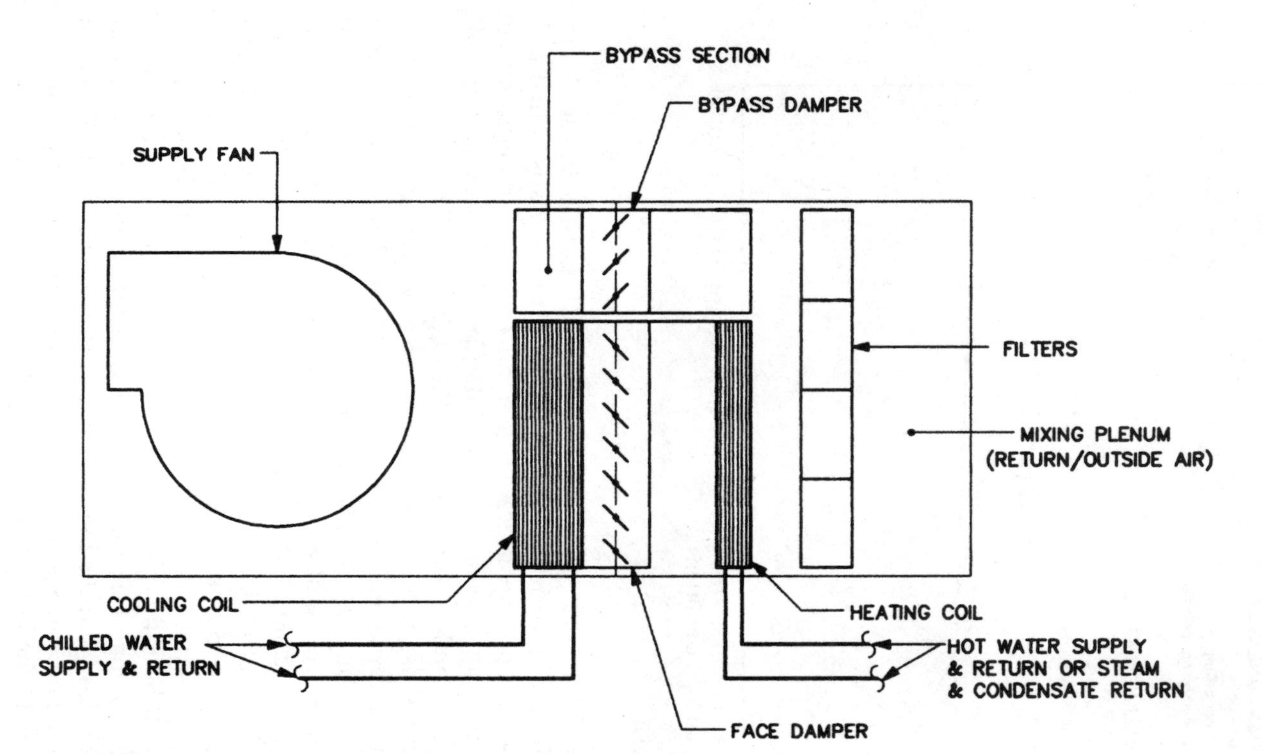

Figure 6.2. Single-zone unit with face-and-bypass control.

Air handling units with face-and-bypass control are physically larger than units with coils and modulating control valves. Consequently, this type of unit generally costs more and occupies more room space. There are, however, advantages to using face-and-bypass control. In the cooling application, because the cooling coil is always providing full cooling capability, the colder coil surface can achieve more effective dehumidification. Consequently, where, in addition to temperature control, dehumidification is also desirable, a face-and-bypass system is a better choice over a system with coil control.

Figure 6.3 illustrates this effect on the psychrometric chart. For a face-and-bypass system under partial load conditions, the leaving condition from the cooling coil is basically the same as that under full load (in reality, the coil leaving condition will be slightly colder because less air is being cooled by the same amount of cooling medium). The air leaving the air handling unit is a mixture of air passing through the face and bypass dampers, which occurs along the straight line between the coil leaving and bypass conditions on the psychrometric chart.

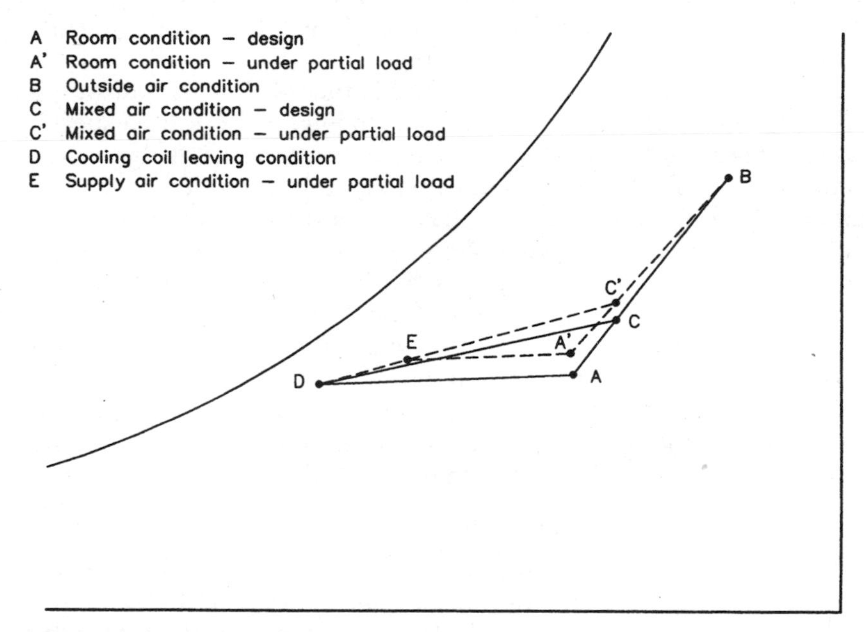

Figure 6.3. Psychrometric plot of a single-zone unit with face-and-bypass control under partial load conditions.

Note that the air leaving the air handling unit in Figure 6.3 has a lower humidity ratio than the equivalent condition in Figure 6.1, which explains why face-and-bypass control offers better dehumidification compared with the coil control.

Face-and-bypass control is also desirable for heating in cold climates. Since there is always full flow of hot water or steam through the heating coil, it offers better protection against freezing of the heating coil.

Control of Single Zone Systems: Reheat

Coil control and face-and-bypass control are the two conventional methods generally used for temperature control of single-zone air handling systems. Since heating and cooling should not occur simultaneously in a single-zone system, it does not matter whether the heating coil is placed upstream or downstream of the cooling coil. In a conventional design, the heating coil for a single-zone air handling unit is generally placed upstream of the cooling coil. In colder climates, this arrangement is necessary to protect the cooling coil, which is inactive during the heating season, from freezing.

Sometimes, however, the heating coil is placed downstream of the cooling coil in a "reheat" position, as shown in Figure 6.4, for a specific purpose. Figure 6.5 shows the psychrometric process of this reheat arrangement. Consider an air-conditioning system serving an auditorium or a meeting room that may have a very low sensible heat ratio (high latent load due to a high concentration of people). As represented by the dashed lines in Figure 6.5, a conventional cooling system will cause the relative humidity in the conditioned space to rise above the comfort condition. In order to maintain the comfort condition in the space, conditioned air can be "overcooled" to a lower temperature to wring out the extra moisture, and with a heating coil placed downstream of the cooling coil, the air can be heated again to satisfy the space temperature requirement, as shown by the solid lines in Figure 6.5.

In this case, the cooling coil is generally controlled by a thermostat placed between the cooling and heating coils or by a dew point controller in the supply airstream that modulates the amount of chilled water supplied to the cooling coil. The room thermostat controls the heating coil so that the supply air to the conditioned space is tempered

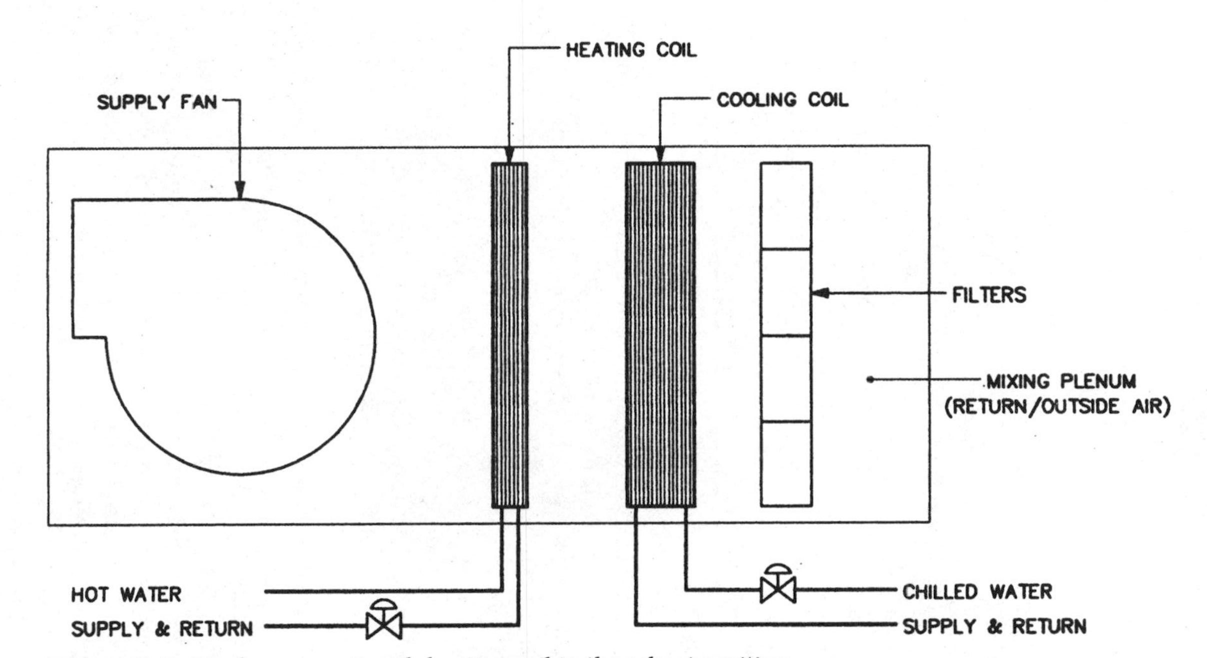

Figure 6.4. Single-zone unit with heating coil in the reheat position.

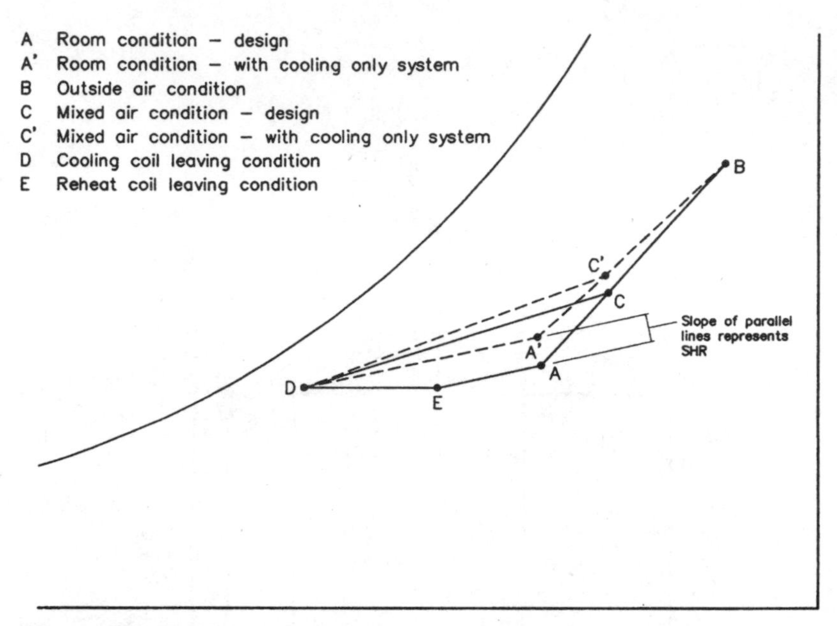

Figure 6.5. Psychrometric process of a reheat system.

by the varying amount of heat applied to the cold airstream. This type of arrangement is referred to as a *reheat system.*

Another way to control reheat is to add a room humidistat together with the room thermostat. Under normal condition, the room thermostat will control the cooling coil to maintain the room temperature as long as the humidistat is satisfied. When the relative humidity in the room rises above the set point of the humidistat, it will override the cooling coil control to supply full cooling. The room thermostat under this condition will activate the reheat coil to maintain room temperature.

One must recognize that the use of a reheat system for this purpose increases the supply air quantity and the cooling energy requirements. Reheat systems also require heating energy in a basically cooling environment. These extra energy expenditures are the price one pays to maintain a specific comfort condition. Consequently, if a conditioned space does not have a specific humidity control requirement, using a reheat system is considered an energy waste and should always be discouraged. On the other hand, if the humidity level in the conditioned space is critical and must be maintained, the extra cooling and heating and additional supply air quantity are all part of the system requirements and cannot be avoided.

7 Categories of Central Air Handling Systems

Constant Volume Systems

As expressed by Eq. (2.1), there are two basic types of air handling systems, *constant volume* and *variable volume*, that can be used to satisfy the load Q_S in a conditioned space. The older of the two groups of central air handling systems is the constant volume system.

By definition, constant volume systems maintain the supply air volume CFM to each conditioned space constant and rely on varying the temperature difference DT (i.e., changing supply air temperature) to meet the load requirements in different conditioned spaces. There are two schemes used in constant volume systems to modify the supply air temperature. One scheme is referred to as a *terminal reheat system,* and the other is called a *mixed air,* or *double-duct, system.*

For the terminal reheat system, the total supply air is cooled such that the supply air temperature from the central air handling unit is cold enough to satisfy the worst condition of all conditioned spaces. When full cooling is not required, the cold supply air is heated up to a higher temperature by a heating coil in the terminal device that supplies conditioned air to the space to satisfy the space needs. While this scheme is very similar to the reheat control discussed in Chapter 6, use of reheat here is different because the primary purpose of terminal reheat is to control temperature in a central air handling system serving multiple conditioned spaces, as opposed to humidity control. Obviously, a side benefit of a terminal reheat system is that in addition to the proper temperature control at each conditioned space, the humidity level is more evenly maintained in all conditioned spaces.

With the mixed air system, the central air handling unit provides two airstreams at different temperatures. Generally, one airstream passes through a cooling coil to supply cold air and another airstream passes through a heating coil to supply hot air. The temperatures of both cold and hot airstreams are designed to satisfy the worst conditions of all conditioned spaces. For each conditioned space, the space temperature is maintained by mixing proper amounts of cold and hot air to neutralize the load requirements. This system is also similar to face-and–bypass control for a single-zone system. Again, the purpose for the face-and-bypass control in a single-zone system is to attempt to lower the relative humidity in the conditioned space, whereas the central mixed-air system is a design scheme primarily for temperature control. Central mixed air systems are commonly referred to as *double-duct*, or *multizone, systems.*

Advantages of Constant Volume Systems

Generally, constant volume systems are simple to design and easy and reliable to operate, with less complexity in control. Before the late 1960s, almost all central air handling systems were constant volume.

These systems, with their less complex control schemes, are particularly suitable for buildings where constant pressure relationships between conditioned spaces are critical. Constant volume systems also have the distinct advantage of having the ability to supply any constant amount of air into any conditioned space to satisfy ventilation needs. While one cannot point at variable volume systems as the culprit of indoor air quality (IAQ) concern, at least we recognize that the concerns are much less severe and easier to deal with in a constant volume system. Hence there has been a conflict between the two major issues, energy conservation versus IAQ, in the selection of air handling systems. Perhaps only time will tell the outcome of the final developments with regard to this major conflict.

Disadvantages of Constant Volume Systems

The major drawback of the constant volume systems is that they generally use more energy. Before the public awareness of the energy shortage, people recognized that buildings required energy to provide a controlled environment and increase productivity. Energy usage in air-conditioning systems was considered an essential part of a building's functional requirements. Since the late 1960s, however, HVAC engineers have begun to explore various design concepts to minimize energy use in air handling systems. Variable volume systems gradually emerged as newer, more energy conserving systems for building air-conditioning. Since the 1973–1974 Arab oil embargo, there has been a tendency to replace constant volume systems with variable volume systems as a better and more efficient allocation of resources.

Since a majority of conditioned spaces do not operate at their peak load conditions for any prolonged periods of time, an air-conditioning system almost always operates under partial load conditions. For a constant volume system to satisfy partial load conditions, the system will either reheat a cold supply airstream, in the case of a reheat system, or mix cold and hot supply airstreams, in the case of a mixed-air system, to obtain proper supply air temperatures needed for different conditioned spaces. The reheat or mixing processes waste both cooling and heating energy.

There are ways to minimize the thermal energy waste in constant volume systems. Furthermore, not all constant volume systems waste the same amount of energy. While a constant volume reheat system may waste a large amount of thermal energy, there are other constant volume systems that do not waste any thermal energy at all.

Constant volume systems also use more *transportation energy,* which is the energy required to move conditioned air through the air handling system. In a constant volume system, by definition, supply and return air fans must operate at a constant air delivery condition at all times; there is no means of saving any electrical energy from the fan motors in a constant volume system. In the past when energy was plentiful and energy conservation was seldom an issue, many air handling systems were designed using high-velocity, high-pressure duct systems to save construction costs and to minimize ceiling space in order to shorten floor-to-floor height or make occupied areas more spacious. These high-velocity systems always incorporate considerably higher fan motor horsepowers compared with their low-velocity, low-pressure counterparts. Energy wastage in these high-velocity systems due to higher motor horsepower is much more pronounced in constant volume systems when compared with variable volume systems.

Variable Volume Systems

Variable volume is not a new concept. It was tried as early as the 1930s and 1940s for both heating and cooling control with little success. The "modern" variable volume system concept for cooling was developed in the late 1960s but gained popularity in the mid-1970s when energy conservation became a serious concern. Since most constant volume systems used more energy to operate, as the energy conservation issue intensified, variable volume systems became so popular that by the 1980s they almost became synonymous with energy conservation. Most of the constant volume systems were viewed as energy wasting designs and targeted as "energy retrofit candidates."

Variable volume systems operate on the principle of modulating the amount of supply air, rather than varying the supply air temperature, to each conditioned space. By definition, the supply air temperature in a true variable volume system is held constant. This implies that a variable volume system can either be a cooling system or a heating system but not both. Since the great majority of load variations in a building relate to the cooling load, it is natural to design variable volume systems as cooling systems. Heating, for a pure variable volume system, is considered separately and generally is not part of the central air handling system. There are exceptions to this basic concept. Chapter 23 discusses a variable volume system that performs both heating and cooling.

Advantages of Variable Volume Systems

Compared with a constant volume design, theoretically, a "pure" variable volume system does not waste any thermal energy. In a variable volume system with a constant supply air temperature, the system responds to the load variation by varying the amount of air supplied into each conditioned space. Under partial load conditions, which practically occur at all times, there is no need to reheat a cold airstream or mix hot and cold airstreams to maintain a particular space condition. Consequently, there is no thermal energy waste.

Air handling units for variable volume systems are usually smaller and require less mechanical space. Ductwork for variable volume systems is also smaller in size and has less "crossover" (ducts crossing each other), which generally requires less ceiling space for air distribution.

Variable volume systems, as the name implies, only move the required amount of air to meet the load requirement at any given time. The total supply air of a variable volume system can be sized for the system block load, as compared with the constant volume system, which has to be sized for the sum of the instantaneous peak loads of each conditioned space. The required fan motor horsepower for a variable volume system under full load is somewhat less than that for an equivalent constant volume system. Furthermore, in a variable volume system, as the cooling loads for various conditioned spaces decrease, supply air volume reduces proportionately with the reduction in cooling load. Under partial load conditions, fan motor horsepower reduces significantly as the volume of air is reduced. The transportation energy savings can be particularly significant for high-velocity systems.

Disadvantages of Variable Volume Systems

Unfortunately, it is very seldom that an air handling system can be a "pure" variable volume system in actual application, leaving the variable volume group with its own drawbacks. Fixing *DT* implies that a pure variable volume system can only be a cooling system, since the supply air temperature is fixed and thus has to be lower than the temperature maintained in the conditioned space. Consequently, some form of supplementary heating must be used in conjunction with the variable volume system to offset the heating load, unless, of course, heating is not needed all year-round, such as the case in some tropical climates.

It is more difficult to maintain pressure relationships between adjacent conditioned spaces in a variable volume system. If a pressure relationship needs to be maintained between a critical area and its surroundings, more complex controls will have to be installed for a variable volume system to maintain that relationship. In a variable volume system, greater consideration has to be given to ensuring that the proper amount of ventilation air (outside air) can be maintained at all times. An overzealous variable volume design, aimed solely toward energy conservation, can impair the minimum air circulation rate needed to maintain required IAQ and comfort conditions in an occupied space. In addition, variable volume systems may not be the best choice for spaces requiring humidity control.

Hybrid Systems

Both constant volume and variable volume systems are important in their own right. Each system group has its own applications, its own merits, and its own shortcomings. Since a pure variable volume system is, practically, a cooling only system, and heating is generally needed for a complete air-conditioning system, one may consider that a pure variable volume system is an incomplete system. On the other hand, constant volume systems have the major drawback of inherited high thermal and transportation energy inefficiencies. Hybrid systems have been developed that combine the advantages of constant and variable volume systems and in turn minimize the disadvantages.

Heating for a pure variable volume system can be accomplished by placing hydronic heating elements or radiant panels along the perimeter of the building. Heating also can be performed by a supplementary constant volume heating (or heating and cooling) system at the building perimeter. More often, however, heating for a variable volume system is accomplished by combining variable volume and constant volume principles into a *common,* or *hybrid, system.* The two commonly used hybrid systems are the variable volume reheat system and the variable volume double-duct system.

Variable volume reheat and variable volume double-duct systems are generally classified as variable volume systems. In reality, however, they are a combination of variable volume and constant volume systems. These systems will operate as pure variable volume systems during most of the cooling season. During heating or low-cooling-load periods, however, they will operate much the same as a constant volume system, using reheat or mixed-air concepts.

The operation of these systems will have a certain amount of thermal energy waste whenever heating and cooling are required simultaneously. Thermal energy saving in these systems is achieved during the period when the system is operating in the variable volume mode. During the reheat or mixing stage, the system inefficiency is essentially the same as that of the constant volume counterpart. However, since the supply air quantity is significantly reduced, the amount of thermal energy wasted is reduced accordingly.

Choices of Central Air Handling Systems

With the hybrid systems added to the constant volume and variable volume system groups, there are many air handling systems to choose from when an HVAC engineer sets out to select a system for a prospective project. There are engineers who are more comfortable with a particular type of system over others and use it most of the time in the projects they design. Other engineers may have more diversified practices. An engineer's practice may cover a variety of building types, which may require different types of systems, one for each type of building project. Most of the time, the original selections of these systems are based on some type of evaluation between initial and operating costs, supported by practical experience. Unfortunately, a young engineer first starting his or her career in HVAC design generally does not have the privilege of sharing the decision-making process on system selection.

Generally speaking, it is easy to design an air handling system if one knows basic system principles and to estimate the cost of installing the system for a given project. It is also relatively easy to recognize the relative significance of energy inefficiencies among various systems under consideration. It is more difficult, however, to estimate the amount of energy waste of one system over another and then to estimate the energy costs associated with the waste.

Computer Simulation of Air Handling System Performance

To better understand the energy use pattern of various types of air handling systems, computer calculation algorithms and methods have been developed to simulate system performance and estimate energy use for different types of air handling systems. It has become relatively easy today to perform a detailed evaluation comparing various central air handling systems.

While computer algorithms have been developed to simulate the performance of various types of air handling systems, it is important to recognize that there are only a few basic air handling system types, i.e., reheat, mixed-air, and variable volume systems. Consequently, only a few basic algorithms are needed for simulating system performance. However, by combining these basic algorithms and modifying their operational or control features, one can generate many unique systems for unique applications. In reaction to energy conservation needs, many energy simulation programs have been written to predict the energy use of building air-conditioning systems. While vendors of such programs claim periodically that a particular program could simulate a greater number of air-conditioning systems than their competitors, the truth is that many use the same algorithms for simulation with very minor modifications for system variations. For this reason, it is important that an air-conditioning engineer have a clear understanding of design principles behind various air handling systems, and it is even more important for such an engineer to understand the algorithms required for each system simulation than to know how to use the simulation program itself.

PART 2

Constant Volume Systems

8 The Terminal Reheat Systems

The constant volume terminal reheat system is one of the oldest central air handling systems developed to serve multiple occupied spaces requiring independent controls. A schematic diagram of such a system with an economizer cycle is shown in Figure 8.1.

In designing any air handling system, after the cooling loads are determined for conditioned spaces, the design engineer generally calculates a set of values for supply air quantities, CFMs, for the conditioned spaces using Eq. (2.2) to satisfy the peak cooling load of the space. By definition, a constant volume terminal reheat system will have to supply the calculated air quantity to all conditioned spaces at all times. In addition, all the supply air will have to be cooled to a sufficiently lower temperature so that when any thermostatically controlled space reaches its design peak cooling load, the temperature of the supply air from the central air handling system can satisfy its need. To meet this requirement, a constant volume terminal reheat system uses a central cooling coil and a supply fan that provides fixed air quantity at a constant temperature. A thermostat is placed at the discharge of the air handling unit controlling the cooling coil capacity to maintain the supply air temperature.

Note that for any given time, there will only be a few spaces that will reach their cooling design peaks. The remainder of the spaces will all be operating under a partial load condition, which can be satisfied with supply air temperatures higher than that supplied from the central air handling system. Since all the supply air has to be cooled down to *DT* degrees (see Eq. 2.2) below the room temperature to satisfy the

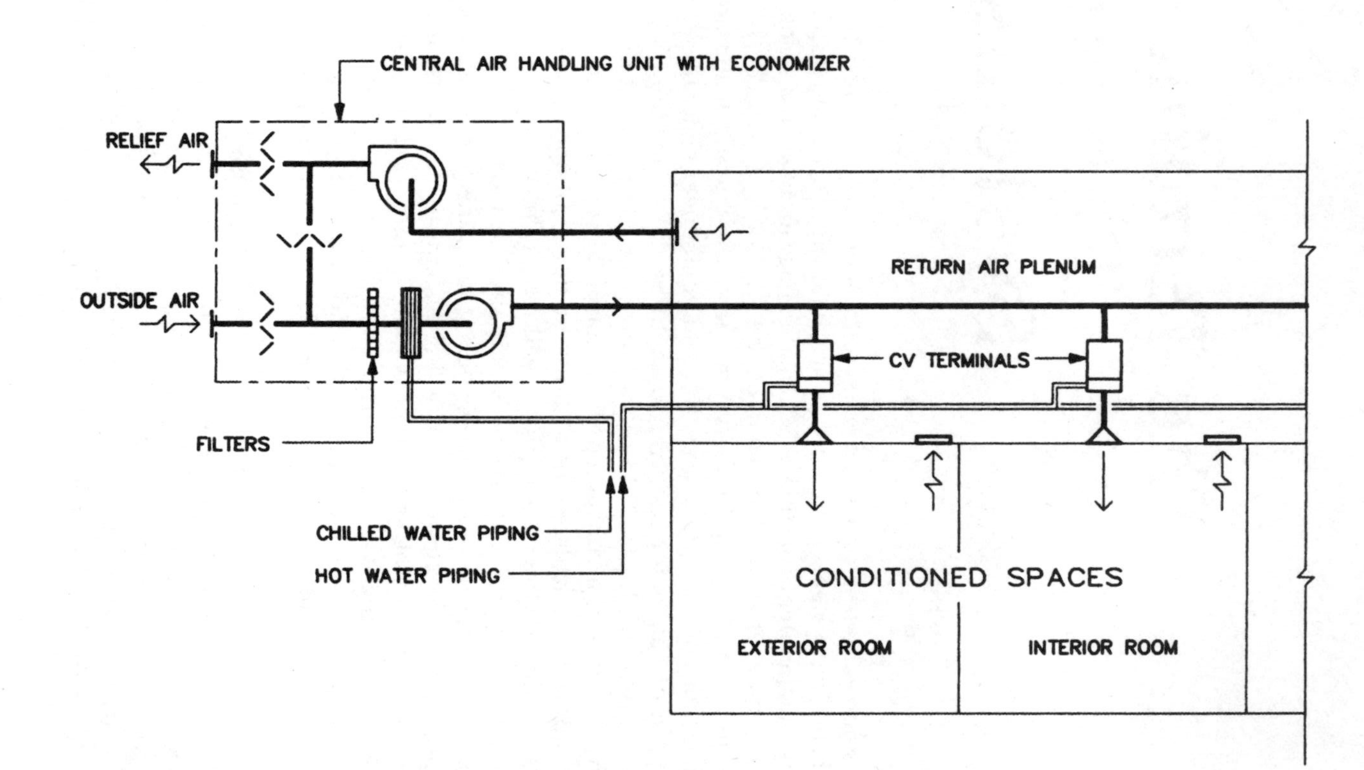

Figure 8.1. Constant volume terminal reheat system.

worst-case space condition, the refrigeration load will have to be higher than what the building needs at any given time.

For all spaces that do not require full cooling, supplying the constant amounts of cold air without any modification to the supply air temperature will overcool these spaces. To satisfy a space that does not need full cooling, a heating source is added in the branch duct serving the space to "reheat" the supply air up to a higher temperature. For each space requiring independent temperature control, a room thermostat will control a heating coil or an electric heater located in the branch duct, usually in a terminal unit serving the space, to modify the supply air temperature to satisfy the room needs. Note that all spaces under partial load condition will require varying degrees of reheat, which represents additional heating energy otherwise not required under a cooling mode. This added heating energy is considered as energy waste necessary to make the constant volume terminal reheat system functional.

The constant volume terminal reheat system was a commonly used system in the 1940s through early 1960s. At the time, not only was energy plentiful, but utility companies also encouraged customers to use energy for better comfort. However, once energy conservation became an important issue, very few of these constant volume terminal reheat systems were suitably efficient to be used for ordinary air-conditioning. Today, new and existing buildings using constant volume terminal reheat systems are often designed or retrofitted with control schemes that will, instead of fixing the supply air temperature from the air handling unit, use load-analyzing logic to reset the supply air temperature higher to satisfy the space requiring the most cooling during partial cooling conditions.

The basic elements in an air handling unit for a constant volume terminal reheat system are filters, cooling coils, and a supply fan, as shown in Figure 8.1. If the system requires 100 percent outside air, a preheat coil is typically needed to heat the outside air up to the supply air temperature, unless the system is installed in a tropical climate where winter design temperatures never drop below the cold supply air temperature. For systems requiring a large percentage of outside air and where the winter design outside air temperature is low, the mixed air temperature of the return air and outside air may be lower than the design supply air temperature. Under such conditions, a preheat coil also may be needed in the mixed air plenum to heat the air up to the cold supply air temperature to reduce the extra burden that will otherwise be imposed on the individual reheat coils.

Constant volume terminal reheat is an effective system for dehumidification control. All the conditioned air passes through the cooling coil and is dehumidified by the cooling process. With reheat control, the humidity level in the conditioned space can be controlled within a reasonable range. A psychrometric plot of a typical constant volume terminal reheat system is shown in Figure 8.2 to illustrate the effect. For applications where year-round humidity control is required, a humidifier can be installed in the air handling unit or in the main supply air duct to add moisture to the supply air during dry winter conditions.

For systems using return air and a minimum amount of outside air to satisfy ventilation requirements, an economizer cycle should be considered so that more outside air can be used for cooling to conserve energy when outside air temperature and humidity are favorable. See Chapter 5 for a more detailed discussion. With an economizer cycle in a system requiring humidity control, the cost savings from using less cooling energy with the economizer cycle should be compared with the additional cost of humidifying larger amounts of outside air (see Chaps. 5 and 32).

The main supply air duct extends from the air handling unit to serve

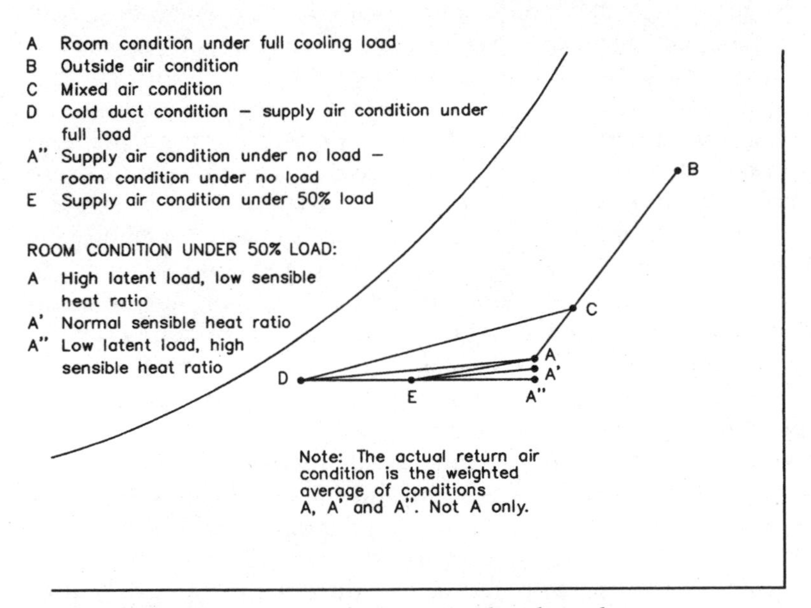

Figure 8.2. Arrangements of reheat at a duct branch.

each functional space requiring air-conditioning. The supply fan and the main supply air duct of the system must be sized to deliver the sum of instantaneous peak air quantities required for all conditioned spaces. Factory-manufactured constant volume reheat terminals with a hot-water heating coil or electric duct heaters are generally used at each branch serving a thermostatically controlled area. The area can be a single room or a group of rooms that face the same exposure and have similar functional uses. Each terminal is sized for the maximum design cooling load. Each reheat coil must be sized to first heat the air from the cold supply air temperature to room temperature and then to heat it to a higher temperature to satisfy the heating load.

It should be noted that in cases where reheat for dehumidification is required, the air quantity for the reheat terminal must be sized for reheat duty under peak cooling design conditions. Reheat terminals are designed in such a way that regardless of the upstream pressure, the terminal will deliver a preset amount of supply air through the branch. These terminals generally are acoustically lined to serve as sound attenuators also.

Factory-manufactured constant volume reheat terminals were not the original design. In the early days, and still in use in some instances today, a simple manual volume damper was installed at each reheat branch for air balance. The room thermostat controlled a heating coil located downstream of the manual volume damper, as shown in Figure 8.3. With this arrangement, once the initial air balance is completed, each branch duct will carry its proportionate share of air quantity to satisfy the peak design load of the area it serves.

The problem with this simple design is that at the air handling unit the pressure drop across the filters increases as the filters perform their function, getting loaded with dirt. The cooling coil will, depending on the condition of the entering air, operate dry or wet. The pressure drop across a wet coil is considerably higher than that across a dry coil. These operational variations affect the performance of the supply fan, and as a result, the amount of supply air will vary as the static pressure across the fan varies.

If the supply fan is adjusted to deliver the required air quantity with clean filters and a dry cooling coil, the system will be short of supply air when the filters get dirty and/or the cooling coil becomes wet. On the other hand, if the supply fan is adjusted to deliver the design air quantity with dirty filters and a wet coil, the system will have too much supply air when the filter is clean and/or the cooling coil is dry. Under these conditions, there will be considerably more wasted

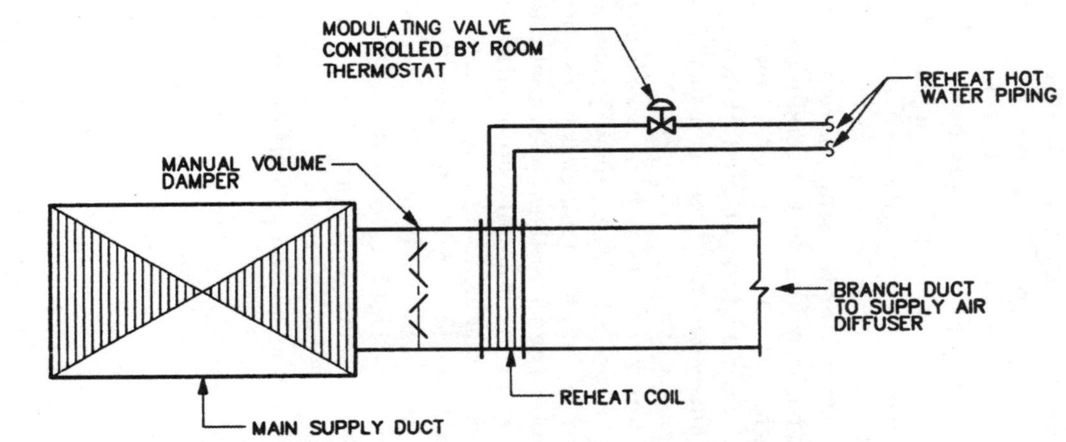

Figure 8.3. Psychrometric process of a constant volume terminal reheat system.

energy because additional reheat energy will be needed to temper the extra air and prevent overcooling of the conditioned spaces. More important, the "constant volume" system will not be operating as a constant volume system under these circumstances.

This problem can be corrected in an existing system in which constant volume control is important and manual volume dampers are used at reheat branches. Automatic control that responds to the system pressure fluctuations can be added to the supply fan. A controller with a pressure sensor located in the supply main can control the supply fan to deliver a constant amount of air at all times. The fan control can be a mechanical device such as an outlet damper, inlet vanes, or a moveable disk to vary the effective width of the fan wheel, or it can be a variable pitch control of a vaneaxial fan or a variable speed drive for the fan motor. As the pressure drop across the air handling system increases, the pressure sensor will sense the drop in static pressure and open up the mechanical control device or speed up the supply fan to maintain the required supply air pressure. As the pressure drop across the air handling unit decreases, the pressure sensor will sense an increase in duct pressure and modulate the mechanical control device toward the closed position or slow down the fan speed.

Sometimes, for better control and maintenance convenience, supply air volume can be calculated and displayed for monitoring purposes. To decide what type of fan control is best suited for a particular project, the engineer needs to consider installation cost, maintenance cost, and energy cost to compare different fan control schemes. One should keep in mind that if the system does not have to operate under constant volume, fan control may not be necessary. Many existing systems operate under this premise.

A constant volume terminal reheat system wastes large amounts of cooling and heating energy and should be discouraged where the only concern is to maintain individual space temperatures. Furthermore, since the system has to deliver a constant amount of air to all conditioned space at all times, the system cannot offer any savings in transportation energy. However, as mentioned in Chapter 6, reheat is also a method of controlling the humidity level in the conditioned space. For buildings where maintaining a specific humidity level is important, such as art museums or auditoriums, a constant volume terminal reheat system is a convenient solution to prevent the humidity level in the conditioned space from rising above the design limit.

Compared with other types of constant volume systems, the terminal reheat system offers the best temperature and humidity control.

The system is very simple to design, easy to install, and requires a relatively small fan room for the air handling equipment and less ceiling space for distribution.

The major disadvantage of the system is that it uses the highest amount of cooling and heating energy. Although commonly referred to as one of the *all-air systems,* this system is really an air-water system, which, in addition to the sheetmetal ductwork distribution, also requires a piping distribution. In the case of electric heat, it also requires an electrical distribution, which involves a different trade. With hot-water reheat, there is a potential water leakage problem with the hot-water piping system in a ceiling space. Because the system involves more than one sheetmetal trade, it is relatively more difficult to make postinstallation modifications.

9 Double-Duct Systems

Constant volume double-duct and constant volume terminal reheat are the two major design concepts of the centralized constant volume systems. Unlike the terminal reheat system, which requires only one main supply air duct, the double-duct system features two main supply air ducts, one carrying cold air and the other carrying hot air. Temperature control for each conditioned space is achieved by mixing varying amounts of hot and cold air from each duct to obtain the proper supply air temperature to neutralize the load requirement. A schematic diagram of a constant volume double-duct system with an economizer cycle is shown in Figure 9.1.

The basic design concept of a double-duct system requires that the cold air must be cold enough to satisfy the worst cooling zone and that the hot air must be hot enough to satisfy the worst heating zone. With the two extreme conditions satisfied, all other partial load conditions can be satisfied by mixing proper amounts of air from each airstream.

Energy waste occurs in the constant volume double-duct system any time the hot and cold airstreams must be mixed to satisfy loads that require neither extreme temperature. Under these partial load conditions, the extra energy needed to cool and heat the air to their design temperatures is wasted. Fortunately, peak cooling and peak heating generally do not occur at the same time. In the cooling season, heating is rarely needed, so there is no need to energize the heating coil. Under these conditions, the mixture of return air and outside air passes through the deactivated heating coil without being heated. For temperature control, a proper amount of cold air can be mixed with the return/outside air to satisfy the cooling load in the conditioned space. Since the heating coil is not activated, there is practically no thermal energy waste for a double-duct system operating in the cooling season.

Many of the cooling load components, such as solar loads and all

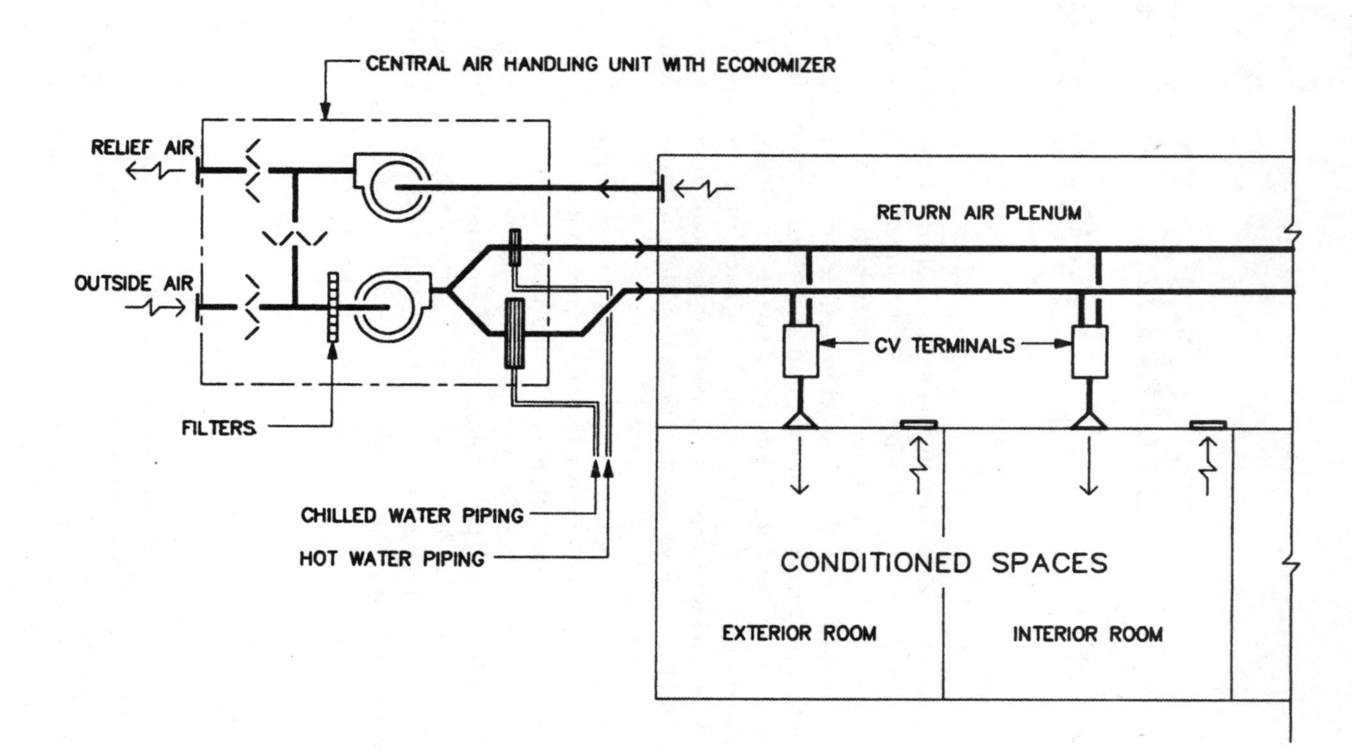

Figure 9.1. A constant volume double-duct system.

internal loads, are not a function of outside air temperature. The cooling load for interior zones can peak at any time. Consequently, the design cold supply air temperature should be maintained at all times. All the heating load components, on the other hand, are a linear function of the difference between indoor and outdoor temperatures. Peak heating load will occur only when the outside air temperature drops to the winter design temperature, and when the outside air temperature approaches the indoor design temperature, the heating load diminishes to zero. For this reason, and to minimize energy waste, the supply air temperature in the hot duct is always reset inversely proportional to the outside air temperature. The hot air temperature is at its design value only when outside air temperature is at the winter design temperature, and the heating is shut off when outside air temperature approaches space design temperature.

Constant volume double-duct systems waste less thermal energy than the equivalent terminal reheat systems. Thermal energy waste is limited to the heating season, when both cooling and heating have to be activated at the same time. The system practically does not waste any thermal energy in the cooling season. Since the system has to deliver a constant amount of air to all conditioned spaces at all times, however, the system cannot offer any savings in transportation energy.

The basic air handling unit serving a constant volume double-duct system consists of filters, a supply fan, a cooling coil, and a heating coil, as shown in Figure 9.1. For a system that requires 100 percent outside air, a preheat coil may be needed to heat the outside air up to the cold supply air temperature unless the system is installed in a mild or tropical climate where winter design temperature is high. The system may or may not incorporate an economizer cycle, largely depending on the ratio of areas between interior and exterior spaces. See later part of this chapter for a more detailed discussion.

The hot and cold main air supply ducts extend from the air handling unit to branch ducts serving each functional space that requires independent temperature control. At each branch, a factory-manufactured double-duct terminal is installed to serve a thermostatically controlled area, which can be a single room or a group of rooms facing same exposure and having similar functional uses. The terminal, responding to a signal from the thermostat, will mix proper amounts of hot and cold air to maintain design room temperature. Manufactured constant volume double-duct terminals are designed with pressure controls that supply a constant amount of air to the conditioned space regardless of the upstream pressure variations in either the hot or the cold air duct. The terminals generally are acoustically lined to serve as sound attenuators also.

Factory-manufactured constant volume double-duct terminals were not the original design. In the early days, and still in use in some instances today, a set of shop-fabricated mixing dampers was installed in the hot and cold branch ducts before they merge into one branch supply duct to the outlets, as shown in Figure 9.2. A room thermostat would modulate the positions of the hot and cold air dampers to satisfy the space needs. A manual volume damper generally was installed at each supply duct downstream of the mixing dampers to facilitate air balance. Once the initial air balance was achieved, each branch duct would get its proportionate share of air quantity from either the hot or the cold duct to cope with the peak design loads of the space it served.

With the field-installed mixing dampers, the same problem of how to maintain a constant supply of air as occurs with the constant volume terminal reheat system also applies to the double-duct system. The pressure fluctuations due to the loading of filters and the wet and dry conditions of the cooling coil affect the air quantity delivered at any given time to the conditioned space.

With the double-duct system, an additional variable enters into the volume-fluctuation problem: Static pressures in the hot and cold air ducts will vary with the heating and cooling demands. For example, static pressure in the hot air duct will be relatively high in the cooling season because there will be very little demand for airflow through that duct. On the other hand, the pressure in the hot air duct will decrease during the heating season because of the increased demand for hot air. While the exterior spaces go through the seasonal changes, interior spaces basically require cooling at all times. These pressure variations between the hot and cold air ducts will affect the amount of air delivered to each conditioned space. It is a fact that a double-duct system with shop-fabricated, field-installed mixing dampers can satisfy the temperature requirement of the conditioned spaces, but it will *not* operate as a true constant volume system.

There are two ways to maintain constant volume in a factory-manufactured double-duct terminal. In the past, mechanical constant volume control devices were used to control the supply air volume, while the room thermostat controlled the proper mixing of cold and hot air to maintain proper room temperature. While this method is still in use today, a more recently developed scheme is to use reset controllers to

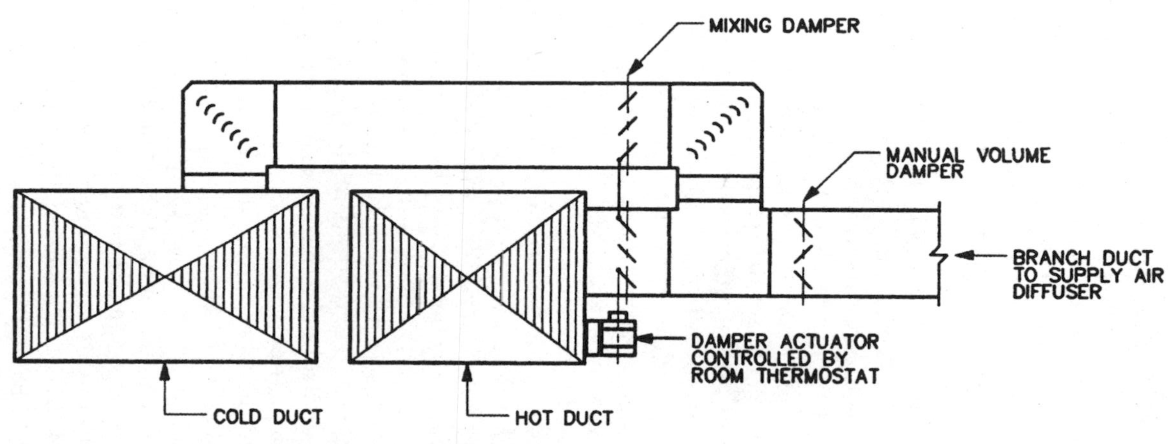

Figure 9.2. Detail of a double-duct mixing damper.

alter cold and hot air damper positions directly to maintain proper mixing temperature and supply air volume at the same time.

Reset controllers use a differential pressure sensor to maintain the required airflow. There are two types of control schemes that employ reset controllers. One is to have differential pressure sensors located in the cold and hot air inlets at the terminal. The controllers respond to the room thermostat signal and position the cold and hot air dampers to maintain proper room temperature. As the pressures in the cold and hot duct fluctuate, the differential pressure sensors through the controller will reposition the dampers to compensate for the pressure changes.

With this control scheme, supply air volume may deviate from constant airflow during the mixing stage. A more positive way to maintain constant airflow is to have one reset controller controlling the cold air damper only. The room thermostat, through the reset controller, modulates cold air damper position to maintain proper room temperature. A differential pressure sensor located in the cold air inlet repositions the cold air damper to compensate for the pressure changes in the cold air duct. Another controller with a differential pressure sensor located in the supply air outlet of the terminal controls the hot air damper to compensate for the varying cold airflows and to maintain constant supply air volume at all times.

The supply fan control, as discussed in the constant volume terminal reheat system, takes care of the pressure variations at the filters and cooling coil but will not be able to cope with the pressure variations between the hot and cold ducts. With this type of design, static pressure control dampers are generally installed in the hot and cold main ducts as they leave the air handling unit in an attempt to minimize the pressure variations. This is particularly true for double-duct systems which utilize shop-fabricated, field-installed damper assemblies.

The temperature-control principle used in a double-duct system—mixing two different temperature airstreams to achieve temperature control—is similar to the face-and-bypass control for a single-zone system discussed in Chapter 6. The primary purpose of mixing air in a double-duct system, however, is for temperature control in a central air handling system serving multiple conditioned spaces, whereas the purpose of the face–and-bypass control in a single-zone system is to lower the relative humidity in a conditioned space while achieving temperature control. Ironically, one of the drawbacks of the double-duct system is that it has an inherently poor dehumidification performance under partial load conditions, particularly when the humidity

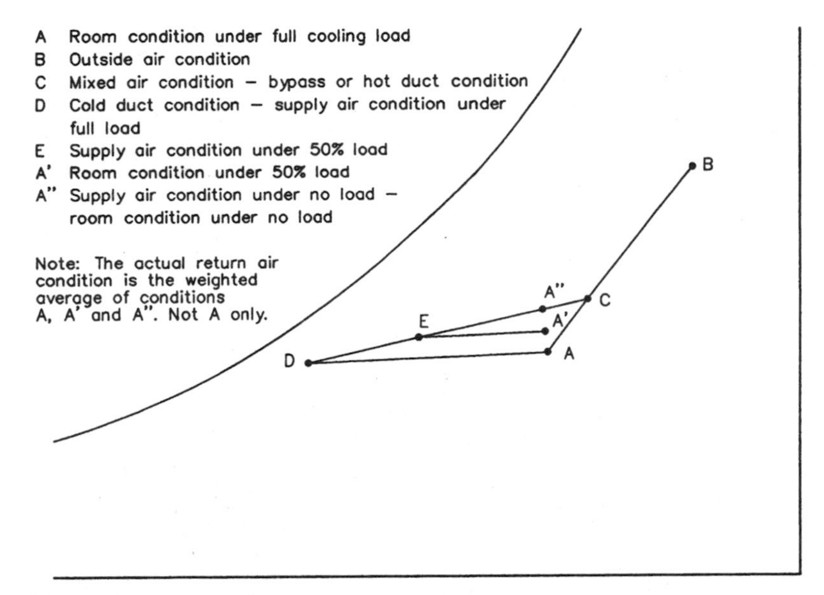

Figure 9.3. Psychrometric process of a double-duct system.

level of the outside air is high. A psychrometric plot of a typical double-duct system is shown in Figure 9.3. Note that under partial load conditions, because the bypass air is not dehumidified by the cooling coil, the conditioned space can have a higher relative humidity if the specific humidity of the outside air is high.

The supply fan of a double-duct system must be sized to deliver the sum of the instantaneous peak air quantities required by each conditioned space. To satisfy the space cooling loads, the main cold air duct needs to be sized only for the air quantity required to match the block load of all spaces served by the air handling system. However, under peak cooling conditions, since the temperature of the air in the hot air duct with a deactivated heating coil is generally higher than that of room air because of higher temperature of the outside air and the heat to return air, some additional quantity of cold air must be supplied to overcome the additional loads.

The reason for installing an economizer cycle in an air handling system is to utilize more outside air to reduce cooling energy when the outside air temperature is low. In a double-duct system, however, since the supply fan delivers the same total quantity of air through both heating and cooling coils, using more cold outside air means that

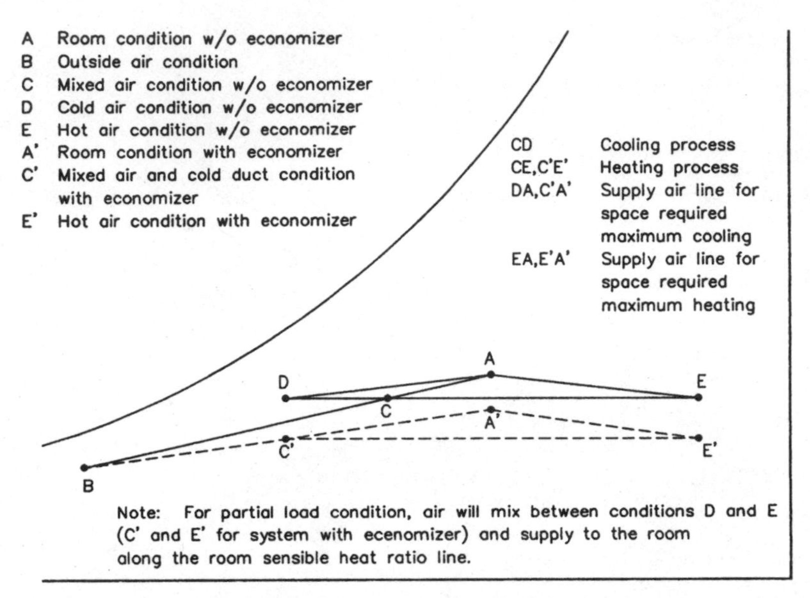

Figure 9.4. Psychrometric process of a double-duct system under winter conditions.

the heating coil has to use more heating energy, since the return air at higher temperature has been replaced by colder outside air. For a system with large interior spaces and minimal exterior exposure, the advantages of using an economizer cycle generally outweigh the extra heating energy required. In a building having primarily exterior spaces with high external loads through windows and walls, however, an economizer cycle can become a burden to the heating system that may exceed any potential energy savings.

It should be noted also that if the system serves spaces requiring humidity control, the cost savings from using less cooling energy with an economizer cycle should be weighed against the additional costs of humidifying the larger amount of outside air. Figure 9.4 shows a psychrometric plot of a double-duct system operating under winter conditions with and without an economizer cycle. Note that the polygon of the psychrometric process will seek its own equilibrium and that with an economizer cycle, the relative humidity in the conditioned space will be lower unless humidifiers are used.

For a system that requires a large amount of outside air and is located where the winter design outside air temperature is low, the mixed air temperature of return air and outside air may be lower than the design cold air temperature. In this situation, a preheat coil also may be needed to heat the mixed air up to the cold supply air temperature. A humidifier also may be installed in the air handling unit if maintaining a minimum year-round humidity level in the conditioned space is required.

The cold supply air of a double-duct system is generally maintained at a predetermined temperature. The quantity of air required for each conditioned space is determined by the cooling load using Eq. (2.2). A hot air temperature needs to be established for the double-duct system so that the heating load of each conditioned space can be met with the room air supplied at this hot air temperature. Using the quantity of air, CFM, established for cooling and the calculated heating load Q_h, the theoretical hot air temperature *TH* required for each conditioned space can be calculated as

$$TH = \frac{Q_h}{1.1 \times \text{CFM}} + TR \tag{9.1}$$

where *TR* is the design room temperature. The space that requires the highest temperature TH establishes the minimum air temperature in the hot air duct. Generally, the hot air temperature is governed by exterior rooms with minimum solar heat gain.

Cooling loads for every space served by an air handling system will not reach individual design peaks at the same time. For any space that is not operating at its full design cooling load, 100 percent cold air is not needed. In order to maintain a constant supply air quantity to a conditioned space under partial load conditions, a part of the supply air will always come from the hot air duct.

Theoretically, the cold air duct can be sized based on the block sensible cooling load of the system, i.e., the maximum load required if the entire area served by the air handling system were one big space. However, since the bypass air carries with it a portion of outside air load and the heat to return air, the amount of cold supply air must be increased to cope with these loads that bypassed the normal cooling process.

Knowing the design temperature difference *DT* for cooling and the mixed air temperature *TM* of outside air and return air, and taking into consideration the temperature rise due to various heat loads to return air, the maximum amount of cold air required under peak cooling load conditions CFMC can be calculated using the following procedure: First, by treating the entire area served by the air handling system as one big space, the required supply air temperature *TSR* to satisfy the block sensible cooling load Q_s for the big space is:

$$TSR = TR - \frac{Q_s}{\text{CFM}_t \times 1.1} \tag{9.2}$$

where TR = design room temperature
CFM_t = total supply air quantity of the constant volume double-duct system

Because of diversity, i.e., not all rooms reach their design peak at the same time, *TSR* will always be higher than the cold supply air temperature *TS*, where $TS = TR - DT$. The proportionate share of total air going through the cooling coil (PCFMC) and, in turn, the cold supply air quantity CFMC can be calculated as

$$\text{PCFMC} = TM - \frac{TSR}{TM - TS} \tag{9.3}$$

and

$$\text{CFMC} = \text{PCFMC} \times \text{CFM}_t \tag{9.4}$$

Note that in Eq. (9.3), PCFMC can be lowered by lowering *TS*. Often, in the interest of reducing cooling coil size and cold air duct size, the amount of cold supply air CFMC is lowered by reducing supply air temperature. One must keep in mind, however, that lowering the supply air temperature will increase thermal energy waste unless a proper control strategy is employed.

The main hot air duct can be sized by establishing air-balance equations between the hot and cold ducts using total heating load and with cold air duct at its design supply air temperature *TS*. To provide higher-temperature air for warmup and/or as a safety factor, and in the interest of minimizing the hot air duct size, a higher maximum hot air temperature is often used.

Compared with a terminal reheat system, the double-duct system offers the advantage of being "all air," i.e., requiring no reheating devices, no piping, or no additional electrical distribution systems. In addition, there is no concern about potential water leakage through the ceiling. Tenant modification and retrofitting work is simplified since there is only one sheetmetal trade involved.

A double-duct system, although not considered very energy efficient, uses considerably less cooling and heating energy than an equivalent constant volume terminal reheat system. Where constant volume is a prerequisite and energy conservation is an important consideration, a constant volume double-duct system is often chosen over a terminal reheat system.

One major disadvantage of the double-duct system is that compared with the terminal reheat system, it does not offer good humidity control in the conditioned spaces. As illustrated in Figure 9.3, humidity level in a conditioned space is a function of the cooling load. Relative humidity in the space gets higher as the cooling load in the space decreases. In tropical applications, outside air may have to be precooled and dehumidified before mixing with the return air to control the relative humidity in the conditioned space. Double-duct systems also require larger fan room and more ceiling space, so construction costs are generally higher than for terminal reheat systems.

In the early days of double-duct system development, energy conservation was not a major concern. Some of the double-duct systems were designed with no control at the cooling coil, with the cold supply air allowed to drift to lower temperatures under partial load conditions. This design variation saves installation costs and also reduces dehumidification concerns under partial load conditions. However, as mentioned earlier, lowering the cold supply air temperature unnecessarily wastes more energy.

In some design applications, two sets of filters are required in an air handling system. Often, the higher-efficiency after-filters must be installed downstream of the supply fan. Sometimes, codes and regulations require after-filters to be located downstream of the cooling coil for health care facilities. This requirement presents an unique problem with the double-duct system that is worth discussion.

With a double-duct system, if after-filters must be installed downstream of the cooling coil, another set of after-filters will have to be installed downstream of the heating coil so that all conditioned air is filtered through an after-filter. Since both the heating and the cooling coils are sized for their peak flow conditions, the sum of the air quantities through both coils is always higher than the capacity of the supply fan. Consequently, the total number of after-filters required for a double-duct system with filters located downstream of the coils is always considerably more than if the filters are allowed to be placed upstream of the coils. Another problem associated with locating filters downstream of the cooling coil is that the filters can get wet because the air leaving the cooling coil is often near saturation. After-filters act like a moisture trap and can become saturated with water. Filter life can be shortened because of wetness of the filter media, particularly when the conditions cycle between wet and dry. The resistance of wet filters can be considerably higher than that of dry filters. In addition, wet filters also can turn into a breeding ground for mold, bacteria, and organisms and thus become counterproductive.

For critical applications, such as clean rooms or special surgical suites in hospitals, very high efficiency after-filters are often required to be located at the room air outlets to protect the critical space from possible contaminants in the air distribution system. However, the need for after-filters downstream of the cooling coil in general patient-related areas in a health care facility is questionable. There is no research support for or scientific proof that locating after-filters downstream of the cooling coil is definitively better than if the filters are located upstream of the coil.

10 Double-Duct Systems with Subzone Heating

Double-duct systems with subzone heating are a derivative of the conventional constant volume double-duct system. The air handling system is identical to that in the conventional double-duct system except that there is no heating coil in the hot plenum at the air handling unit. A portion of the supply air passes through a plenum without a heating coil in parallel with the cold plenum. The hot air duct in the conventional double-duct system becomes a bypass air duct in this system. Cold air and bypass air main ducts and the air-mixing terminals for this system are essentially the same as those in the conventional double-duct system. The only difference is that since the bypass main duct has no heating capability, small hot-water heating coils or electric duct heaters must be installed downstream of the air terminals to provide heating for exterior zones where heating is needed.

A subzone heating system is more commonly applied to all-electric buildings with electric duct heaters instead of hot-water heating coils installed at the terminals. The reason is obvious, since electricity is available throughout the building, and the cost of installing a complete hot-water distribution system can be avoided. A schematic diagram of this system is shown in Figure 10.1. Since the bypass air main duct does not carry tempered air, it may not require thermal insulation.

The purpose for this rearrangement of the conventional double-duct system is energy conservation. As stated in the preceding chapter, a double-duct system does not waste thermal energy in the cooling season because the heating coil does not have to be activated. Waste

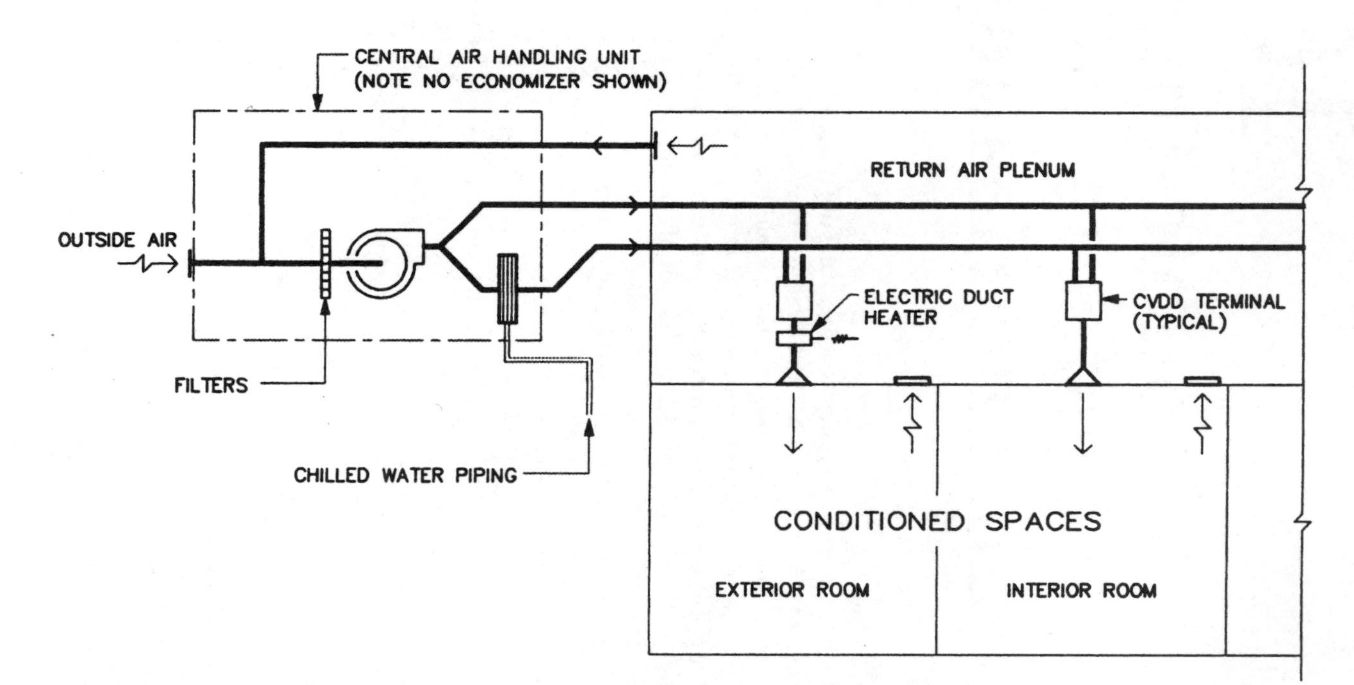

Figure 10.1. Constant volume double-duct system with subzone heating. (Electric duct heaters shown. Terminals may have hot-water heating coils.)

occurs only in the mixing action during the heating season when both heating and cooling coils must be activated at the same time. For interior spaces where cooling is required year-round, there is no need for air in the hot duct to be heated above space temperature. With subzone heating system, since there is no heating coil at the air handling unit, the thermal waste of mixing hot and cold air is eliminated. There will be no energy waste for the interior spaces year-round.

For an exterior space where both heating and cooling are required, the heating will not take place until the cold air damper is completely closed. As the cooling load reduces, the cold air damper will modulate closed and the bypass air damper will modulate open. As the space requires heating, the heater will be activated to cope with the varying heating loads while the cold air damper is closed. Heating the mixture of outside air and return air does not constitute any thermal energy waste.

The use of an economizer cycle in conjunction with this type of system should be carefully examined. Since the economizer cycle can only try to satisfy the cold air temperature by mixing more cold outside air with return air, without a heating coil in the bypass air plenum, the bypass air duct and cold air duct will have the same air temperature. Interior spaces that rely on mixed-air principle to satisfy space needs will lose control under economizer operation unless a tempering coil is installed in the bypass air plenum to heat the air to return air temperature.

Double-duct systems with subzone heating are seldom chosen for buildings requiring 100 percent outside air or a high percentage of outside air unless a heating coil is placed in the bypass air plenum to heat the mixed air or 100 percent outside air to room temperature. The installation costs of a subzone heating system are generally higher than those of a conventional double-duct system. With the added cost of piping and control for the heating coil in the hot plenum, it may even be more difficult to justify the use of subzone heating system.

In the cooling mode, the psychrometric process of a double-duct system with subzone heating is the same as that for conventional double-duct systems, as shown in Figure 9.3. The only difference is that in the heating mode, instead of hot air being heated centrally, the heating action takes place at individual terminals and heat is applied to the bypass air only.

There is an additional feature of a double-duct system with subzone heating. With the heating coil located downstream of the cooling source, the control scheme at each individual zone can be modified to perform reheat for dehumidification if needed. A humidistat can be installed in addition to the thermostat in spaces requiring humidity control. When the humidity in the conditioned space exceeds a preset limit, a control relay can override the normal control of the mixing dampers and allow 100 percent cold air to supply the space. The thermostat, in order to satisfy the temperature requirement of the conditioned space, will activate the heating coil to heat the cold supply air and maintain the desired room temperature. Figure 10.2 shows a schematic control diagram for this arrangement. Figure 10.3 shows the psychrometric plot of this arrangement comparing with conventional double-duct systems which clearly illustrates the effectiveness of this control modification. It is important to note that if reheat for dehumidification is required for a particular space under full load conditions, the supply air quantity for the space must be calculated based on the reheat requirement, which will generally be higher than if the space required temperature control only.

One drawback of subzone heating system is that the heating part of

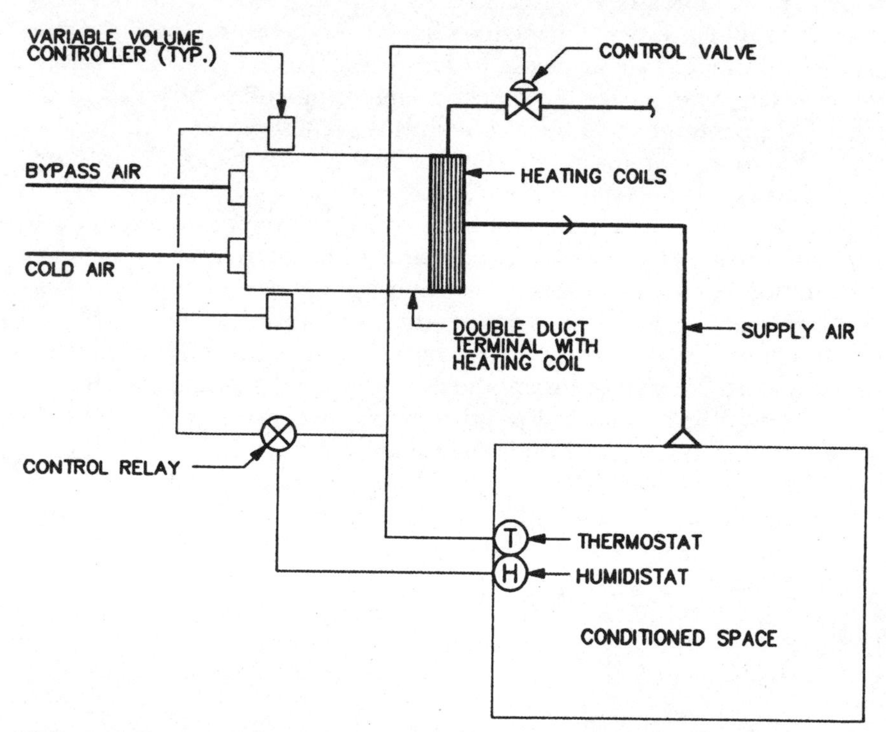

Figure 10.2. A control scheme variation for a subzone heating system.

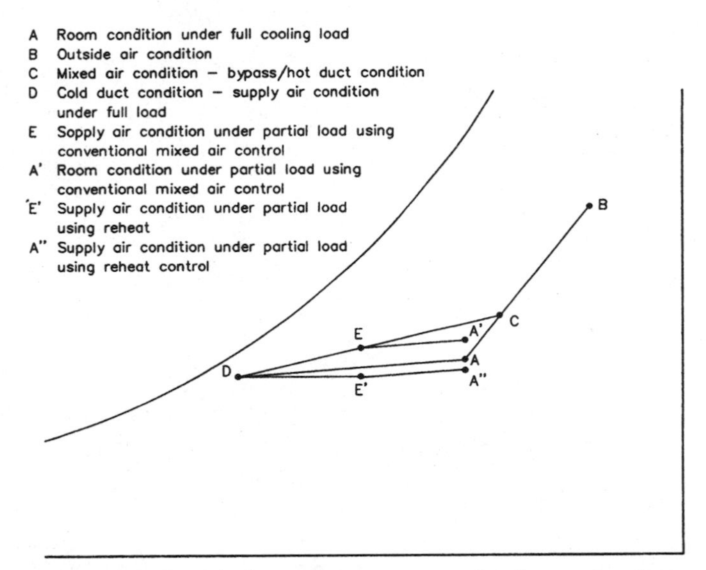

Figure 10.3. Psychrometric process of subzone heating used as reheat for humidity control.

the system is no longer centralized. A series of electric circuits or a hot-water distribution system will have to be provided for those terminals requiring heating, and this involves installation costs that are generally higher than those for conventional double-duct or terminal reheat systems. This is particularly true if an electric heater or hot-water heating coil is still needed in the hot plenum for tempering the supply air through the bypass duct in the heating season. A life-cycle analysis should be made to evaluate whether or not the energy savings resulting from use of a subzone heating system are worthwhile.

11

Double-Duct Systems with Draw-Through Cooling

The double-duct system with draw-through cooling can be viewed as a reheat system in a double-duct format. It also can be viewed as a hybrid system using both reheat and mixed air principles. A schematic diagram of this arrangement with an economizer cycle is shown in Figure 11.1.

With conventional double-duct systems, the mixture of return air and outside air is split after the supply fan and goes through either the cooling coil or the heating coil to be cooled or heated. With the draw-through double-duct system the cooling coil is located upstream of the supply fan; 100 percent of the mixed air will flow through the coil and be cooled to the cold supply air temperature. This cold supply air is then split after the supply fan, with a portion of it going directly into the cold duct and another portion going through a heating coil in the hot plenum which heats the cold air to the hot supply air temperature. The duct distribution and mixing terminals of this system are identical to the conventional double-duct system.

The primary reason for using this type of system is that it offers the same excellent dehumidification characteristics as the terminal reheat system but without the extensive reheat piping and control associated with a single-duct terminal reheat system. The reheat is done centrally rather than at individual terminals. Figure 11.2 shows the psychrometric plot of a typical double-duct with draw-through cooling system. Comparison with the psychrometric plot of the conventional double-

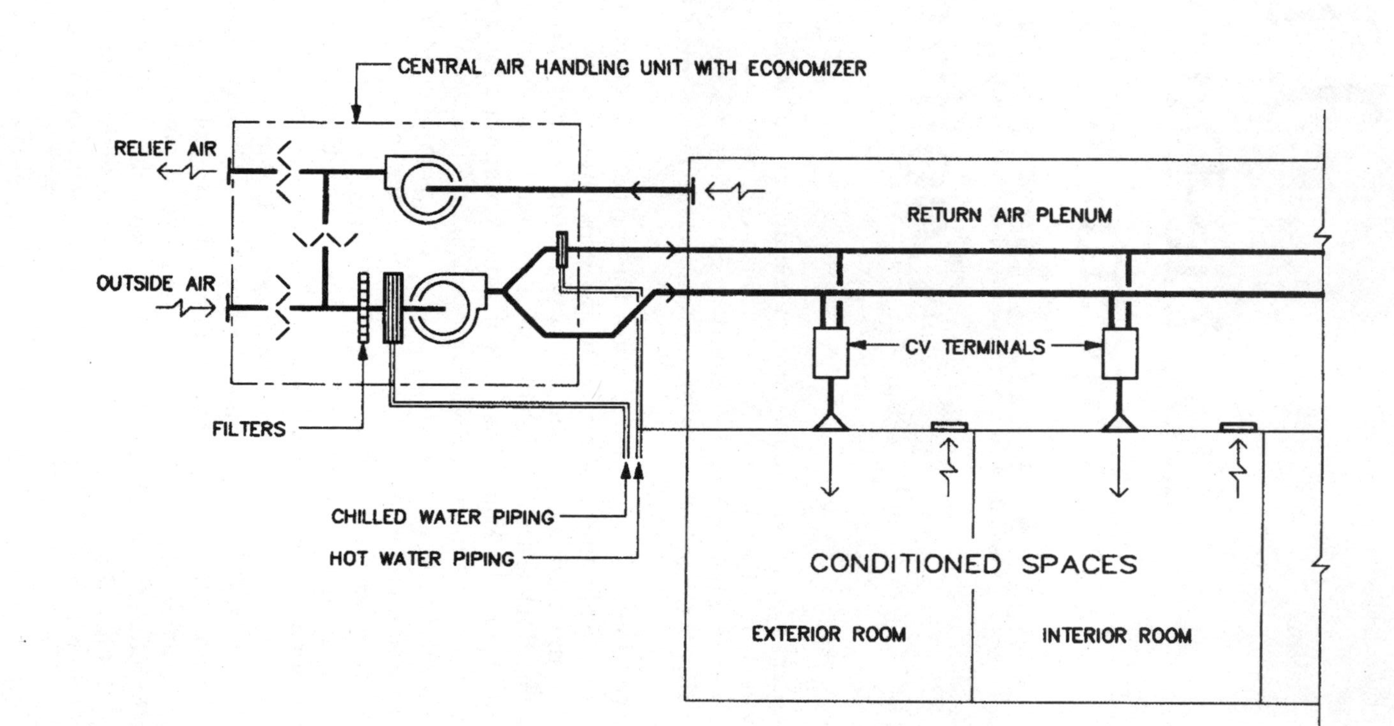

Figure 11.1. A constant volume double-duct system with draw-through cooling.

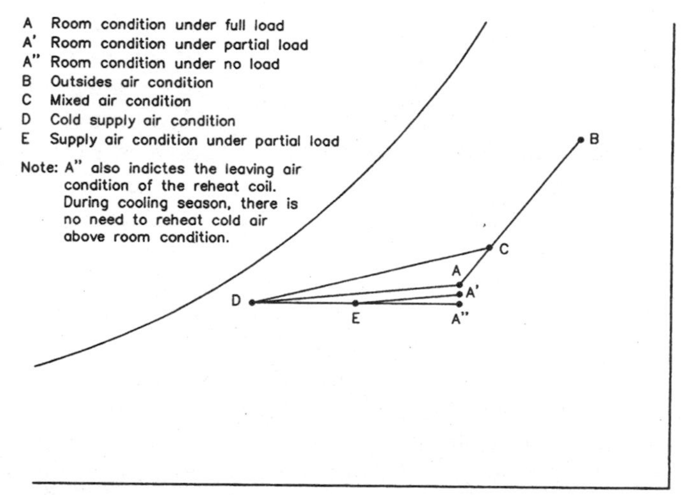

Figure 11.2. Psychrometric process of a double-duct system with draw-through cooling.

duct system (Fig. 9.3) indicates that with this system, both hot and cold air is delivered at the same specific humidity. Consequently, the humidity level in the conditioned space will not be affected by the varying proportions of the hot and cold air, which was the case with the conventional double-duct system. Line *DA*" in Figure 11.2 indicates the amount of reheat needed under cooling conditions for air passing through the hot duct. Note that during the cooling season, heating is still needed to heat the cold supply air to the room condition for the hot duct. During the heating season, the heating coil in the hot plenum will heat the air to a temperature that can satisfy the worst heating zone.

The conventional double-duct system wastes energy for reheating when using an economizer cycle. With the draw–through system, all air passes through the cooling coil, and the heating coil is actually a reheat and heating coil at all times. Since the reheating penalty is there whether the economizer is used or not, an economizer cycle should be used whenever possible to reduce the cooling energy consumption, unless humidification is a part of the requirement, in which case an evaluation should be made to compare the cooling energy savings from use of an economizer versus the extra humidification costs required.

Like all double-duct systems, this system also requires more ceiling space for the distribution system. However, a good feature of this system is that the distribution system of this "reheat" system involves the sheetmetal trade only. There is no reheat piping or additional electrical wiring involved over the occupied space. Modification and retrofitting work is thus simplified. The system is particularly suitable for applications where having reheat piping over the conditioned space is intolerable but dehumidification and humidity control in the conditioned space are important, such as art museums or critical spaces in hospitals.

The major drawbacks of this system are higher construction costs and the fact that the system wastes considerably more energy than the equivalent conventional double-duct system. In the cooling season when no heating is required, the reheat coil in the hot plenum will still have to heat the cold air up to the space temperature so that no space will be overcooled because of no load in the space. From the energy-consumption viewpoint, a double-duct system with draw-through cooling performs much the same as a terminal reheat system. Application of this system type is limited to buildings where humidity control is a major concern and where reheat piping cannot be tolerated over the conditioned space.

12 Two-Fan Double-Duct Systems

The two-fan double-duct system is an elaborately modified double-duct system designed specifically to maximize the efficiency of thermal energy use. To achieve this goal, however, the air handling unit becomes more cumbersome and complex, to the point that two air handling units are used for one system. While the two-fan double-duct concept can be applied to a constant volume system, it is more commonly used in the variable air volume application (see Chap. 28). There are many pros and cons of this system that deserve detailed discussion.

The duct distribution and mixing terminals of this system are identical to those of the conventional double-duct system. The only difference occurs at the air handling unit. A schematic diagram of a two-fan double-duct system is shown in Figure 12.1.

The original reason for creating this type of system was an attempt to stop the waste of heating energy when an economizer cycle is used in a conventional double-duct system. In a conventional double-duct system, the supply fan delivers the same mixture of return and outside air through the cooling coil and the heating coil. During cold weather, the economizer will deliver mixed air at a temperature to satisfy the cold air requirement. More heating energy will be required for the heating coil because otherwise, without economizer cycle, the temperature of the air entering the heating coil can be considerably higher than the temperature of the air in the cold air plenum.

A two-fan double-duct system uses two supply fans, one for cooling and one for heating. An economizer cycle is used with the cooling side

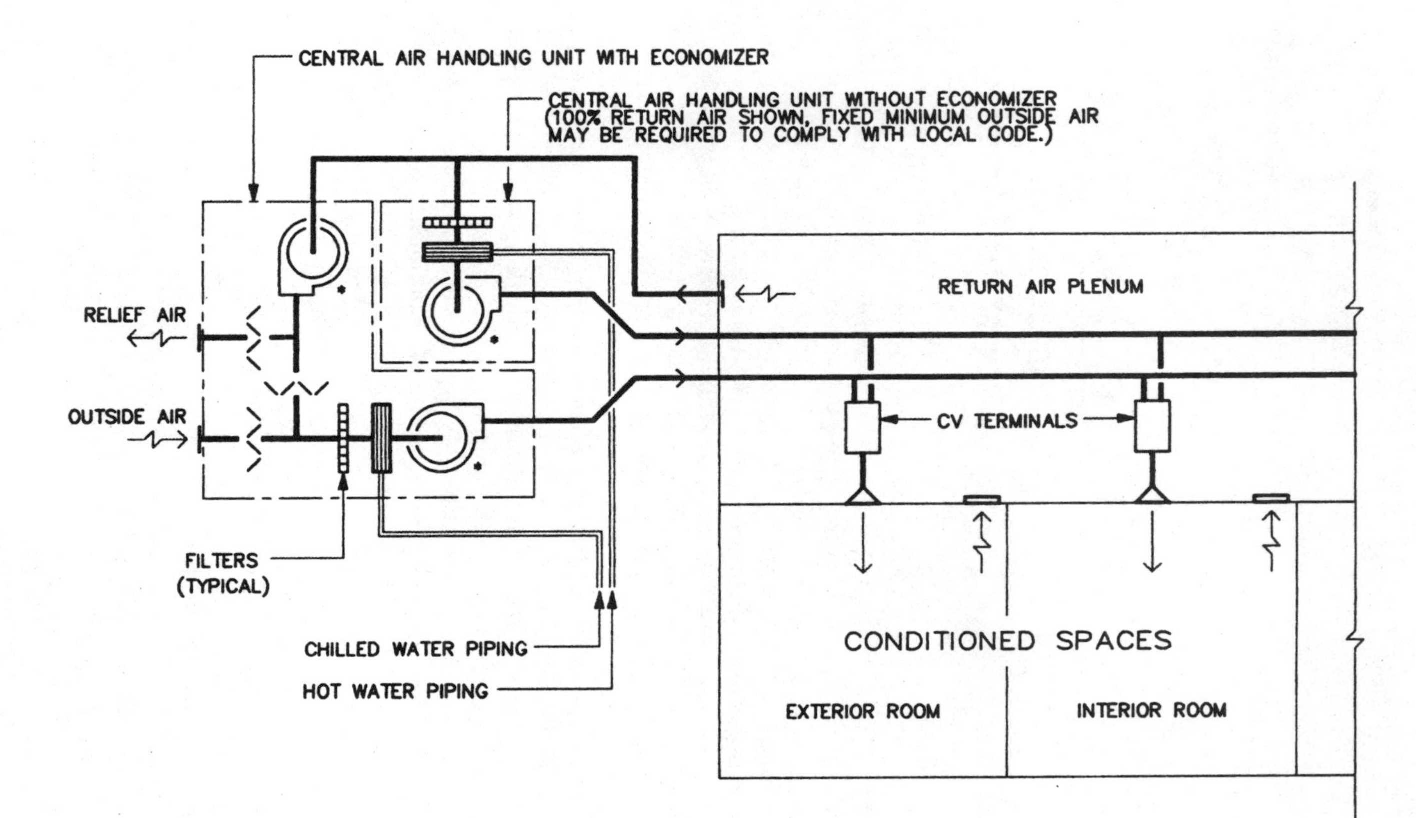

Figure 12.1. A constant volume two-fan double-duct system. (All fans need to have variable volume control.)

of the system. The heating side of the system generally uses 100 percent return air. With the increasing concerns about indoor air quality (IAQ) in recent years, some two-fan double-duct systems have used fixed minimum amounts of outside air on the heating side to satisfy minimum outside air requirements. There is no reason to incorporate an economizer on the heating side since, by definition, an economizer utilizes cooler outside air during mild weather to minimize cooling energy, thus serving no purpose for heating.

With the two-fan double-duct system, the economizer cycle operates on the cooling side only, and the heating side has its own supply fan. The economizer cycle operation of bringing in more cold outside air to the cooling side does not affect the heating side. In fact, the general consensus is that in a two-fan double-duct system, outside air is not needed at all for the heating side, and 100 percent return air can be used for heating to truly minimize the use of heating energy. The rationale for using 100 percent return air on the heating side is that (1) very rarely will a conditioned space require 100 percent hot air, and (2) even so, the only time any conditioned spaces require heating is when the economizer cycle is in operation, and during such periods, there will be more outside air going through the overall air handling system than the system requires for minimum ventilation.

Unfortunately, while the two-fan double-duct system conserves energy, it also has some disadvantages and drawbacks. For example, the system requires two supply fans. Consequently, the fan room for a two-fan double-duct system must be considerably larger than the comparable fan room required for conventional double-duct systems. With added fans, there also will be added maintenance and equipment care.

Not only must the number of fans be increased, but also the combined capacity of these fans must be considerably larger than in the conventional single-fan double-duct system. In a constant volume double-duct system, both the heating coil and the cooling coil are always sized for more than 50 percent of the total air. For example, an average system may have a cooling coil sized for 80 percent of the total supply air quantity and a heating coil sized for 60 percent. At any given time, however, the supply fan of a conventional double-duct system will supply air through the cooling coil and the heating coil proportionally to satisfy the load requirement. By definition, the sum of supply air quantities through both cooling and heating coils will always equal 100 percent of design air quantity. Consequently, the supply fan of a conventional double-duct system can be a constant volume fan and be sized to satisfy the total air quantity only.

With the two-fan system, however, each fan will have to be sized to satisfy the maximum capacity required for each coil. The total fan capacity of a two-fan system, that of the cooling fan plus that of the heating fan, is always larger than the sum of the air quantities required by the conditioned spaces. For the example mentioned above, the cooling fan must be sized for 80 percent of the total supply air quantity, and the heating fan must be sized for 60 percent of the total supply air quantity; thus 140 percent of total fan capacity will be needed for the two-fan system. Together with the fan size increase, the number of filters must increase accordingly. These larger fans add more mechanical equipment to the HVAC system and require larger fan room spaces, both contributing to a significant construction cost increase.

The supply fan of a conventional constant volume double-duct system does not need any variable volume control. With the two-fan double-duct system, however, volume control devices should be used for both cooling and heating fans in order to conserve transportation energy and prevent excessive pressure builtup in the duct system as building loads shift between cooling and heating. More controls add more construction costs, require more calibration, and need more maintenance.

In summary, the use of a two-fan double-duct system for a constant volume application can minimize heating energy waste, but it will have higher initial costs and will be cumbersome to install and maintain. While this system may be suitable for energy conservation in cold climates, a detailed life-cycle analysis pairing the initial expenditure and maintenance concerns to the potential energy savings should be done to justify its use. A variable volume form of this system, however, is usually a more viable and worthwhile solution to energy conservation.

13 Triple-Duct Systems

Triple-duct systems are rarely used anymore, particularly after the variable volume systems became popular in the 1970s. The system was developed in the 1960s primarily for energy conservation when the practical use of variable volume systems was still in doubt. The triple-duct system is discussed here mainly for its historical value and as a parallel discussion to the triple-deck multizone systems discussed in Chapter 16, which still are in practical use. A schematic diagram of a triple-duct system is shown in Figure 13.1.

The air handling unit for the triple-duct system has a bypass plenum in addition to the conventional hot and cold air plenums. The mixture of return air and outside air can either be heated, cooled, or go through the bypass plenum. Three main ducts—hot, cold, and bypass air—will be extended from the air handling unit into the occupied space. The system uses the same conventional constant volume double-duct terminals. For interior spaces that require cooling at all times, the double-duct terminals will be connected to cold and bypass ducts. For exterior spaces that require heating in winter and cooling in summer, the terminals will be connected to hot and cold ducts.

With this system, thermal energy waste during the heating season for the interior zones is eliminated. However, energy waste still occurs for the exterior zones where hot and cold air must be mixed to satisfy the space needs. An economizer cycle should not be used in conjunction with this system, since with the economizer in operation, the interior spaces will lose control because the cold duct and bypass duct will have air of the same temperature. For the same reason, this system is not applicable to air handling systems that require a high percentage of outside air.

The construction costs for a triple-duct system are considerably higher than those for the other constant volume systems. This system

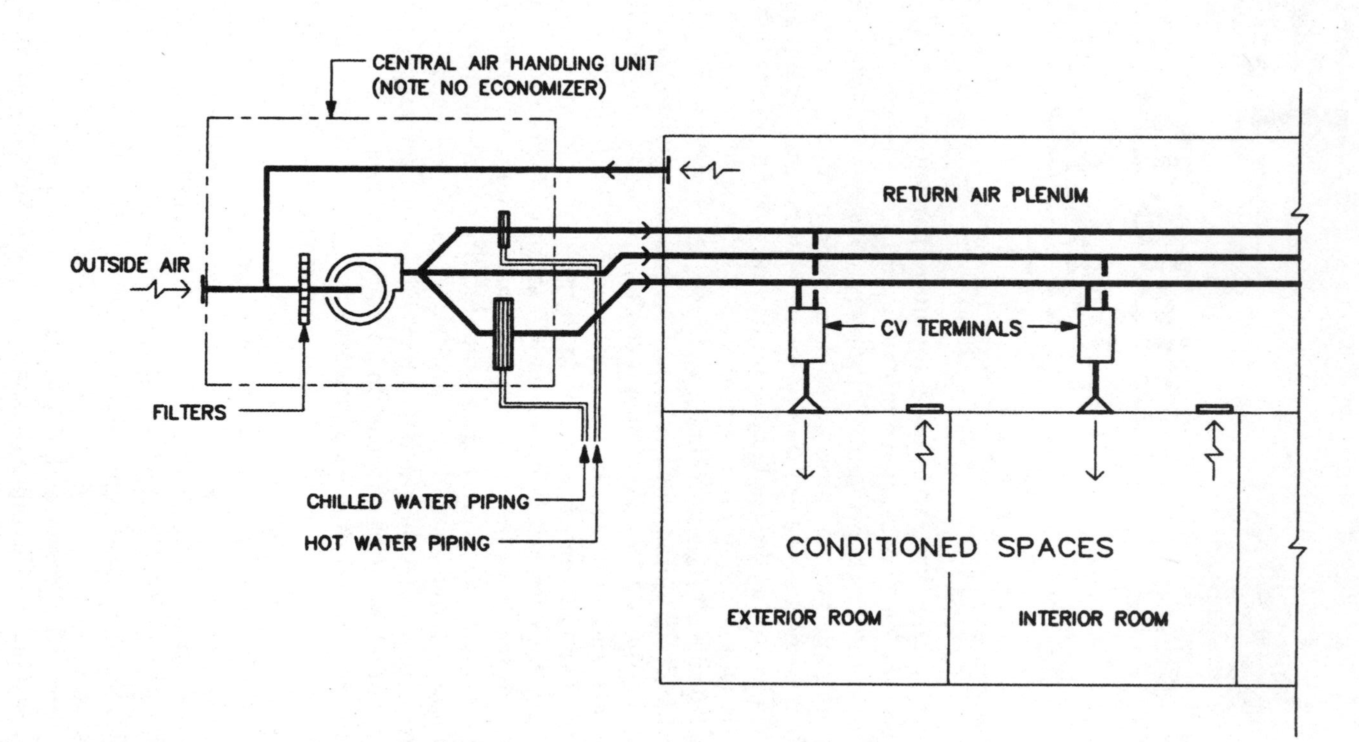

Figure 13.1. A constant volume triple-duct system.

also requires more ceiling space for installation because of the additional bypass duct. With no potential for transportation energy savings, the system cannot be justified unless the extra construction costs can be offset by the savings in thermal energy alone. With the advent of variable volume systems, there is very little reason to consider this system unless "constant volume" is a firm requirement and the thermal energy savings justify use of this system under some unique circumstances.

14 Multizone Systems

A multizone system is a variation of the double-duct system. Parallel to the different types of double-duct systems, multizone systems also have several design variations that will be discussed separately.

Unlike the double-duct system, which has hot and cold ducts extending into the occupied areas with mixing terminals for temperature control located over the conditioned spaces, the multizone system performs temperature control for groups, or "zones," of conditioned spaces at the air handling unit by mixing hot and cold air with individual pairs of control dampers located at the air handling unit. A manual volume damper should always be placed downstream of each pair of mixing dampers to facilitate air balance. A schematic diagram of a typical multizone system is shown in Figure 14.1.

The thermal characteristics of the multizone system are identical to those of the double-duct system. The psychrometric process of a multizone system is also the same as that for the double-duct system (see Fig. 9.3). The problem of lacking dehumidification capability of the constant volume double-duct system is also a drawback of the multizone system.

A multizone system is best suited for serving a relatively small floor area that contains custom-built spaces requiring independent space thermostatic control. The system is very simple to design, requires minimum ceiling space, is easy to install, and has all control elements centrally located at the air handling unit. However, because the duct system from a multizone system is generally tailored to a fixed configuration of the conditioned spaces, flexibility for future tenant modification is limited.

Unlike the constant volume double-duct system, which has constant volume terminals to regulate the amount of air supplied into each con-

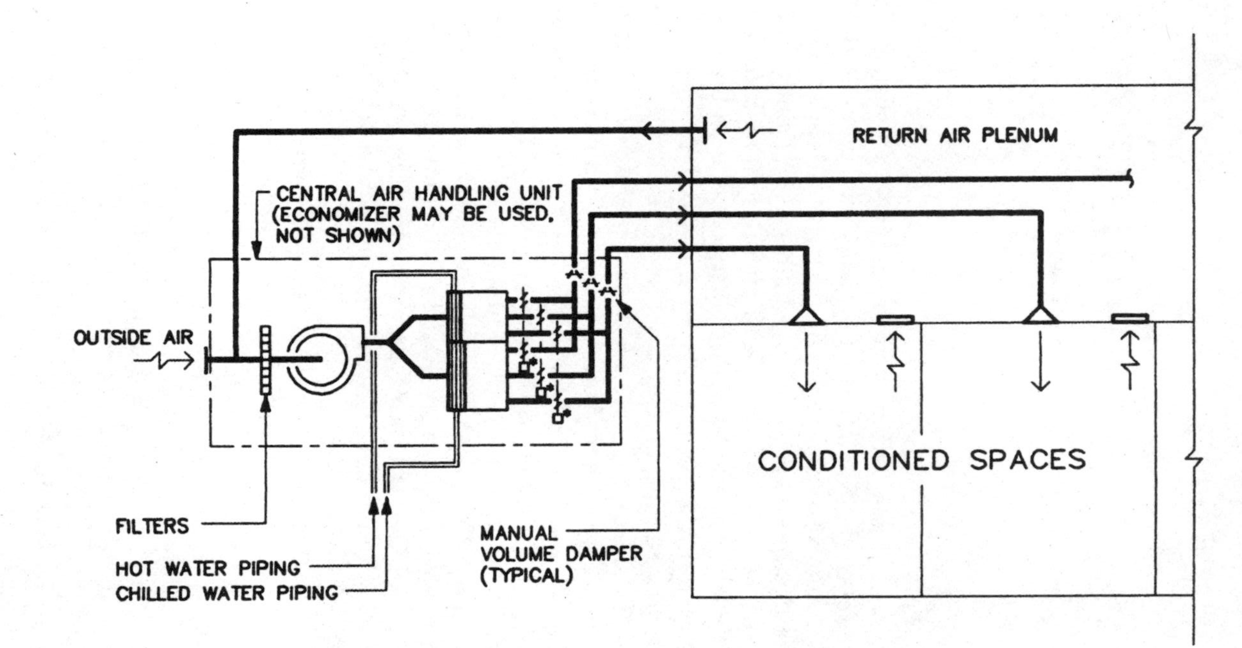

Figure 14.1. A multizone system. (*Mixing dampers; only three zones are shown for clarity.)

ditioned space, the mixing dampers in the multizone unit do not have a constant volume control device. Consequently, as discussed in relation to constant volume double-duct systems, the air quantities delivered to the conditioned spaces will vary as the static pressure fluctuates as a result of filter loading and the wet and dry conditions of the cooling coil. Since multizone systems are customarily used for smaller buildings and often for "open plan" applications, the system does not have to maintain constant volume, and system pressure variations generally do not present an operational problem.

The use of an economizer cycle with a multizone system needs to be examined carefully. Most multizone systems serve a relatively small floor area that has a large exterior-to-interior space ratio. Since the economizer cycle only benefits cooling for interior spaces in the heating season and wastes heating energy for exterior spaces, it is often difficult to justify the use of an economizer cycle for systems serving an area with relatively small interior areas. In some cases, a misapplied economizer cycle can actually use more heating energy than the amount of cooling energy the system is able to save.

Energy waste for a multizone system is also the same as that for a double-duct system. To minimize energy waste, outside air temperature reset is normally used for the hot air plenum temperature control, and the heating coil is generally deactivated during the cooling season. Being a constant volume system, the system cannot offer any savings in transportation energy.

Another disadvantage of the multizone system is that it is generally custom selected and arranged for fixed functional needs. Altering the system to fit a changed room configuration generally involves significant modifications to the existing system or some compromise in space comfort.

15

Multizone Systems with Subzone Heating

Multizone systems with subzone heating are a modification of the conventional multizone system. The relationship between a conventional multizone system and a multizone system with subzone heating is equivalent to that between a conventional double-duct system and a double-duct system with subzone heating. The duct distribution of a multizone system with subzone heating is identical to that of the conventional multizone system. The only difference between these two systems is the location of the heating source in the air handling system.

With this system, there is no heating coil in the hot air plenum, which thus becomes a bypass plenum. For interior spaces, since there is no need for heating year-round, the mixing dampers can mix cold and bypass air to satisfy all space needs. For exterior spaces, a hot-water heating coil or an electric duct heater is added in the branch duct downstream of the mixing dampers. Heating will not be activated until the cold air damper is completely closed and the bypass air damper is fully open. A schematic diagram of a typical multizone system with subzone heating is shown in Figure 15.1.

This modification of the conventional multizone system was designed to conserve energy. There is no energy waste for the air serving interior spaces because the mixing action is between the cold air and the bypass air. Under heating conditions, the cold air damper will always be closed, and all supply air will come through the bypass

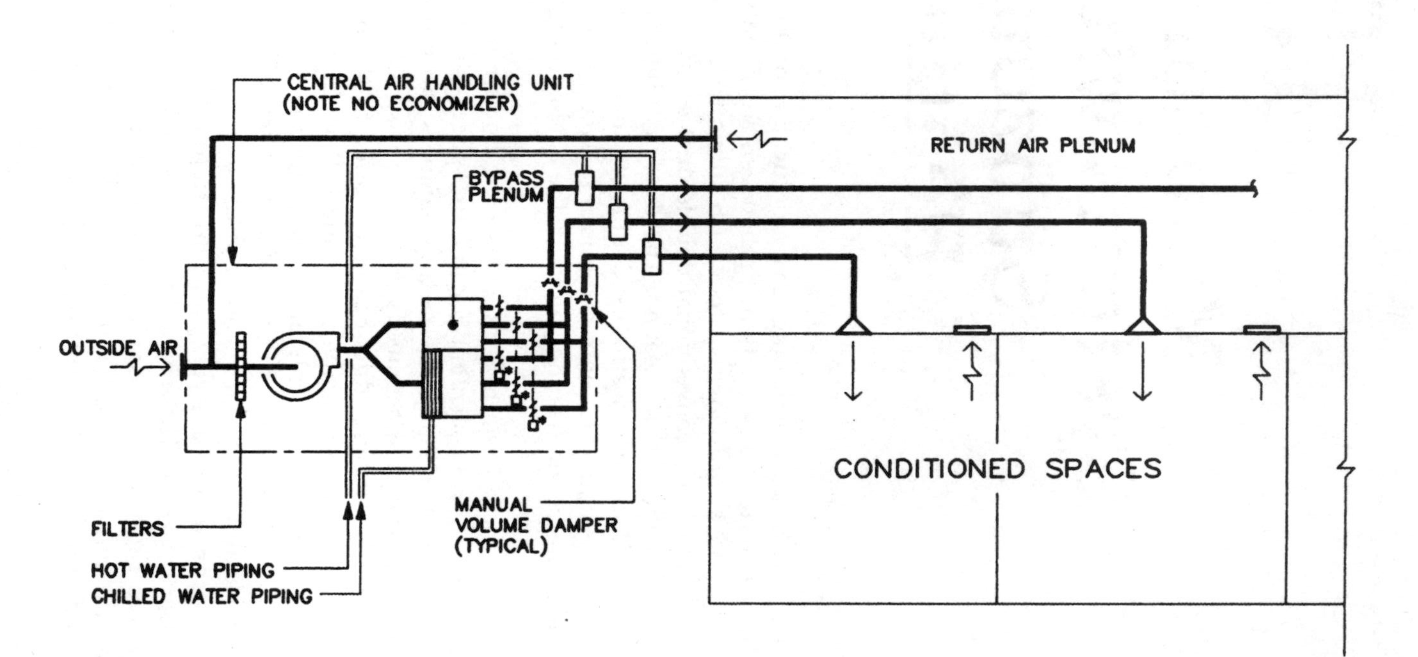

Figure 15.1. A multizone system with subzone heating. (*Mixing dampers; only three zones are shown for clarity.)

damper. Heating the mixture of outside air and return air does not constitute any energy waste, so the system is considered to be thermally energy efficient.

Like the double-duct system with subzone heating, an economizer cycle generally should not be used in conjunction with this system. When the economizer cycle is in operation during cool weather, the economizer cycle control will mix the proper amount of cold outside air with return air to satisfy the required cold air temperature. Without a heating coil in the bypass air plenum, the bypass air and cold air will have the same temperature. Interior spaces, which rely on the mixed-air principle to satisfy space needs, will lose control completely unless another heating coil is installed in the bypass plenum to heat the air up to room temperature. This type of system also should not be used for applications where a high percentage of outside air is required.

The primary reason for using double-duct or multizone systems with subzone heating is energy conservation. A multizone system with subzone heating is a more practical application compared with a double duct system with subzone heating. Subzone heating coils for multizone systems are generally located in the fan room, where they are easier to maintain and repair, whereas the subzone heating coils for double-duct applications are located in the ceiling space above the occupied areas with an extensive hot-water distribution system, which will be considerably more costly and difficult to service and maintain.

Besides the improvement on energy usage and the impracticality of using an economizer cycle, other advantages and disadvantages of multizone systems with subzone heating are basically the same as for conventional multizone systems. Similar to the double-duct system with subzone heating, the multizone system with subzone heating also has the possibility of offering a dehumidification control scheme that is identical to that described for the double-duct system with subzone heating (see Chap. 10).

16 Triple-Deck Multizone Systems

A triple-deck multizone system is very similar to the conventional multizone system with one unique modification at the air handling unit. Similar to the triple-duct system discussed in Chapter 13, a bypass plenum is added at the air handling unit in conjunction with the conventional hot and cold air plenums. The mixture of return air and outside air is split after the supply fan and can either be heated, cooled, or go through the bypass plenum. Multiple sets of mixing dampers, each consisting of a hot, a cold, and a bypass damper, will be controlled by thermostats in the conditioned spaces. A schematic diagram of a typical triple-deck multizone system is shown in Figure 16.1. While triple-deck multizone systems use a similar concept to the rarely used triple-duct system discussed in Chapter 13, the triple-deck multizone system is a very practical and energy-conserving constant volume system.

The bypass plenum in a triple-deck multizone unit acts as a buffer zone in the temperature-control scheme. Hot and cold air in the triple-deck multizone system never mixes together. For a conditioned space demanding full cooling, the cold air damper will open and the bypass and hot air dampers will close. As the cooling load in the conditioned space reduces, the cold air damper will modulate closed and the bypass air damper will modulate open while the hot air damper remains in the closed position. As the cooling load is reduced to zero, the cold air damper will close completely and the bypass air damper will open fully. As the load changes from cooling to heating, the cold air damper will remain closed, and the bypass air damper will modulate with the hot air damper to provide a proper supply air tempera-

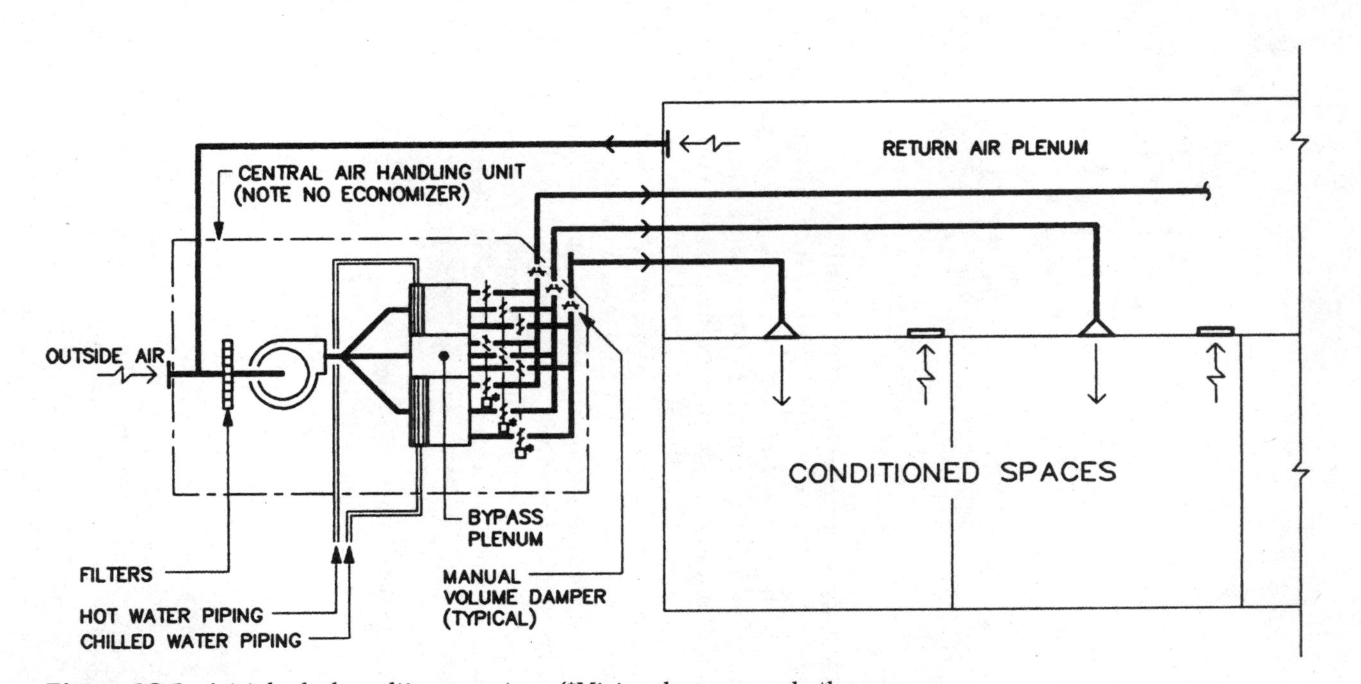

Figure 16.1. A triple-deck multizone system. (*Mixing dampers; only three zones are shown for clarity.)

ture to the conditioned space. Most manufacturers use a specially designed common shaft with linkages to operate all three dampers simultaneously with a single actuator to perform the control function described above. Since the hot and cold airstreams never neutralize each other, there will be no thermal energy waste with this system.

The duct distribution of the triple-deck multizone system is identical to that of the conventional multizone system. For reasons mentioned in the discussion of subzone heating systems, an economizer cycle should not be used in conjunction with this system, and mechanical cooling will have to be used during the heating season. With the use of an economizer cycle, the bypass plenum will have the same temperature as the cold plenum. Interior spaces that rely on modulating the mixed air temperature between the cold air and bypass air will lose control. For the same reason, this system should not be applied to situations where a high percentage of outside air is required.

Both the triple-deck multizone system and the multizone system with subzone heating were designed for energy conservation. Since no individual subzone heating coils are involved, the triple-deck multizone system is simpler to install. This system, however, does not have the dehumidification feature offered by subzone heating systems.

17 Fan Coil Unit Systems

The use of fan coil units for air-conditioning can be viewed as linking together a large group of small single-zone air handling systems each providing air-conditioning to a relatively small area. Each fan coil unit is a very basic air handling system, consisting of a fan, a filter, a heating coil, and a cooling coil. Sometimes, in lieu of a hot-water coil, an electric duct heater is used for heating. Occasionally, only one coil is used for both heating and cooling. The one-coil scheme is used either with a two-pipe system with seasonal changeover, where the piping system carries chilled water in summer and hot water in winter, or with a four-pipe system with changeover control such that the coil can use chilled water or hot water anytime as the load shifts between cooling and heating.

Fan coil units can be used conveniently and efficiently for building air-conditioning. They are particularly suitable for providing air-conditioning to hotel guest rooms, high-rise apartments, and sometimes patient rooms in hospitals.

One of the major drawbacks of using fan coil units is the decentralized maintenance. Since these fan coil units are sometimes located in or above the ceiling space directly over the occupied areas, servicing these units can become a maintenance inconvenience. In the past, multiple fan coil units also were used frequently to provide air-conditioning for large office buildings. With the development of large central air handling systems, however, fan coil units are rarely used for office building air-conditioning today.

Fan coil unit application will require that chilled water, hot water, and electricity be provided to each unit. Where electric heat is used in lieu of hot water for heating, the hot-water distribution system is replaced by a larger electrical distribution. Since cooling often involves dehumidification, moisture will normally condense from the cooling coils, and a drain line will have to be installed for each fan coil unit to drain condensate to the sanitary sewer system. In this regard, some codes also require an overflow secondary drain line or an overflow alarm be installed for each fan coil unit to prevent water damage from accidental blockage of condensate drain lines.

To provide proper ventilation, outside air also has to be supplied to each fan coil unit or to the areas they serve. Outside air can be supplied locally for fan coil units serving exterior spaces where exterior wall openings can be arranged as a part of architectural design. This method of introducing outside air, while marginally acceptable for buildings in mild climates, is highly vulnerable to variable wind conditions and can cause freeze-up problems in cold climates. More often, a centralized air handling system is used to supply outside air to the decentralized fan coil units. In this case, outside air can be filtered, preheated, or precooled as required.

In most instances, the duct distribution of the centralized outside air does not have to be connected directly to each individual fan coil unit. Instead, outside air is often supplied into the ceiling cavity above the occupied space serving as a mixed air plenum for ventilation air and return air at each fan coil unit. Figure 17.1 shows a schematic diagram of a typical fan coil unit application. The companion psychrometric plot is shown in Figure 17.2.

The use of fan coil units for air-conditioning is energy efficient. Not only is there no reheat or mixing waste of thermal energy associated with this system, fan energy is also minimized because the air distribution is localized. The amount of energy required to pump chilled and hot water is considerably less than the fan energy required by centralized air handling systems.

It is impractical to use an economizer cycle with each fan coil unit. Consequently, the advantage of using outside air to cool is lost. However, since most fan coil unit applications are for exterior spaces, the energy savings potential from use of an economizer cycle for these applications is limited.

Fan coil units are particularly convenient for hotel rooms, apartments, and hospital patient rooms, since each of these spaces always has a companion private toilet room that requires exhaust. The required outside air matched with the toilet exhaust keeps the conditioned space in natural air balance. Fan coil units are also suitable for retrofit applications where the building was not air-conditioned previously. Under these circumstances, the building generally does not have sufficient ceiling space for the duct distribution system of a central air handling system.

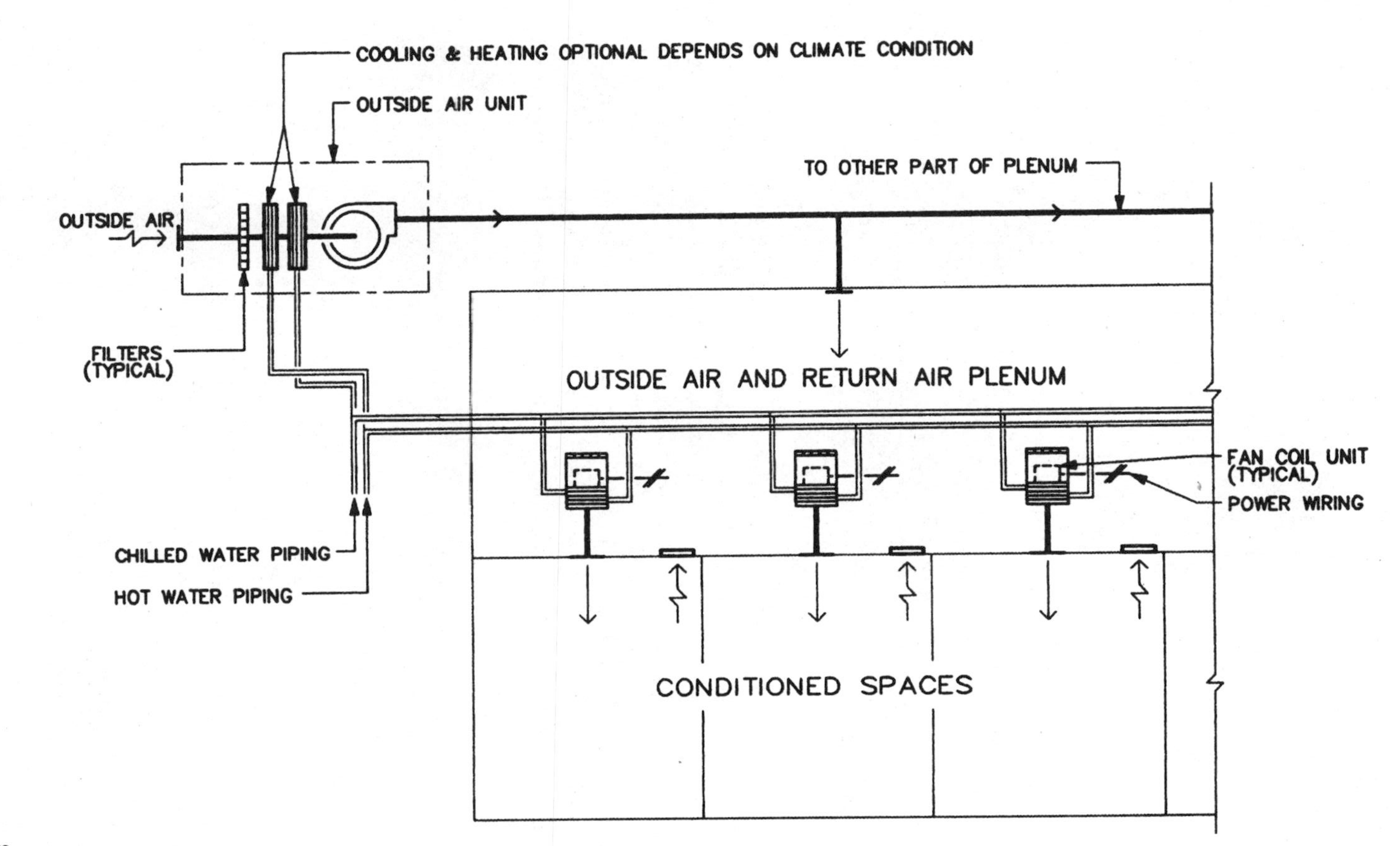

Figure 17.1. Fan coil units.

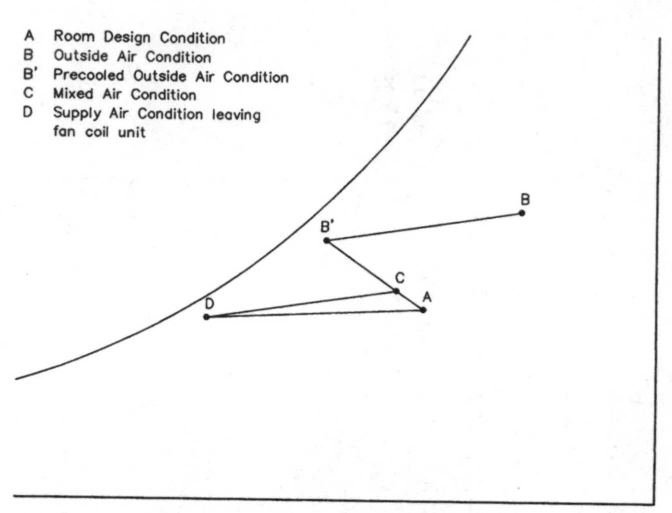

Figure 17.2. Psychrometric process of a fan coil unit system with a centralized outside air system.

The major disadvantage of using fan coils units is the significant added maintenance time for operating personnel and the increased inconvenience to the occupants associated with servicing this system. Sheetmetal, piping, and electrical trades can all be involved in maintenance and repair of these units. Condensate drainage also can become a nuisance. Fan coil units also can develop noise problems because of the small blower and motor located in or directly above the occupied space. For these reasons, fan coil units are seldom used for large office buildings today.

18 Induction Systems

The induction system is a unique air-conditioning system that uses induction principles to circulate air in the conditioned space. An induction system consists of a central air handling system that delivers a relatively small amount of conditioned air, referred to as *primary air,* under high pressure to *induction units* located in the conditioned spaces. In the induction unit, the primary air is discharged through induction nozzles at high velocity, which induces a certain amount of room air, referred to as *secondary air,* and supplies the mixture of primary and secondary air into the conditioned space. A schematic diagram of a typical induction system is shown in Figure 18.1, and two types of commonly used induction terminals are shown in Figure 18.2.

An induction system can be viewed as a variation of a fan coil unit system in which the small blower in a fan coil unit is replaced by primary air and induction nozzles. In other words, the high-pressure centrally conditioned air becomes the prime mover in the induction system to supply further conditioned air into the occupied space.

Although return air can be used in the primary air, in most instances, 100 percent outside air is used for the primary air system. In this case, the primary air system is also equivalent to the centralized outside air system for the fan coil unit system. In this analogy, it should be noted that using high-pressure air as the prime mover of the conditioned air is generally not as energy efficient as using small blowers driven by electric motors. Consequently, the induction system tends to use more transportation energy compared with fan coil unit systems.

Induction units are generally installed under windows along the perimeter of a building. Induction units also can be mounted in the ceiling space, generally behind a soffit of a lowered ceiling area with supply outlet located in the front face of the soffit.

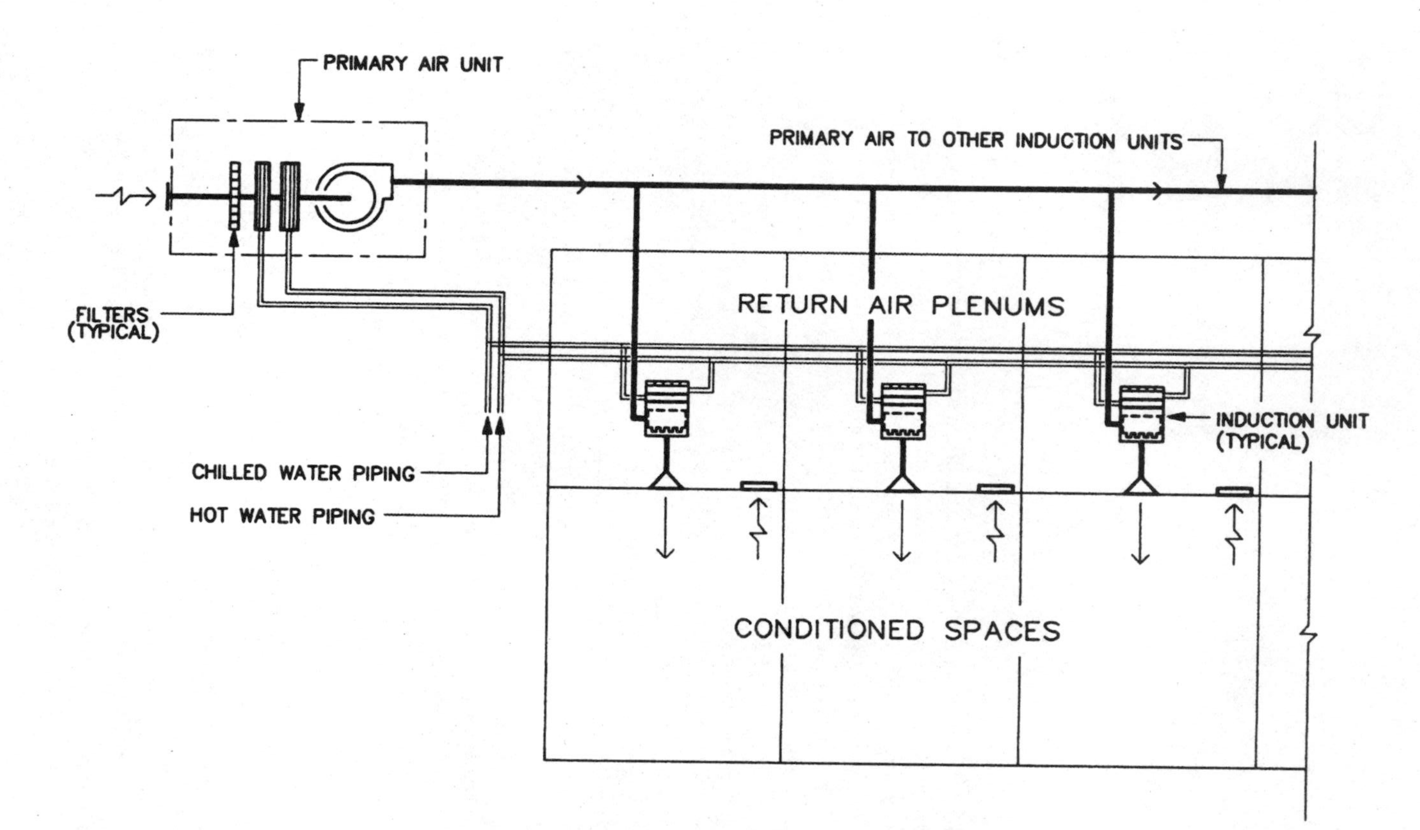

Figure 18.1. An induction system. (Four-pipe system is shown; two-pipe system is similar. Ceiling-mounted induction unit is shown; window sill unit is similar.)

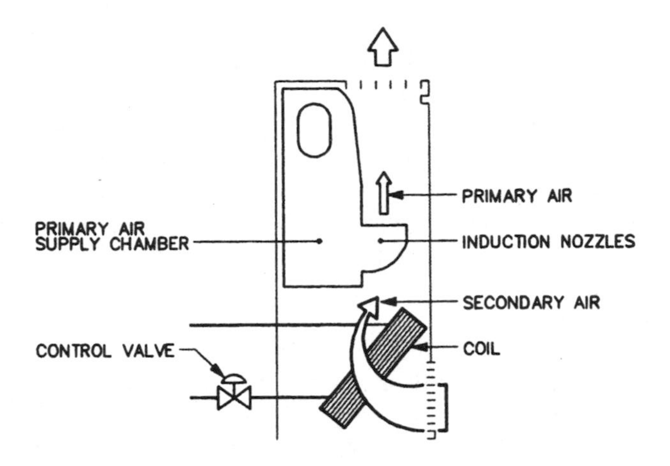

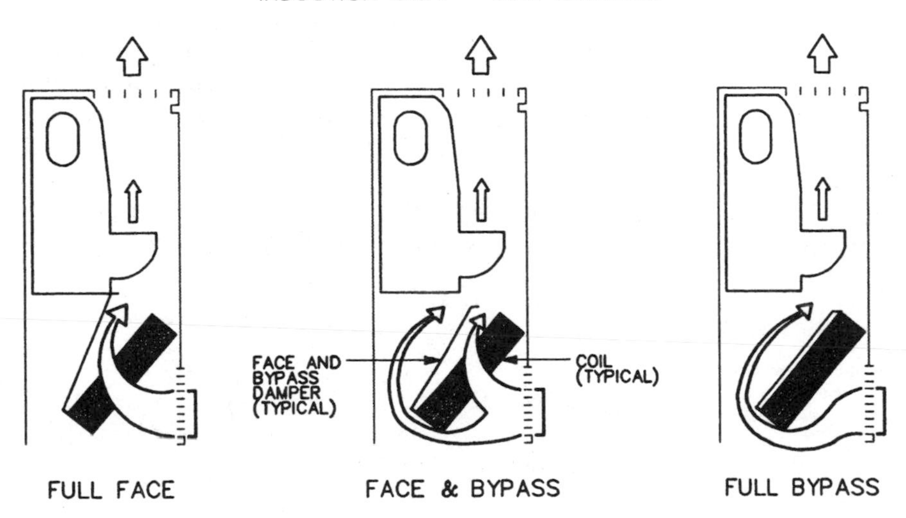

Figure 18.2. Different types of induction units.

There are several different induction system designs. In all cases, however, the primary air is always supplied by a high-pressure air handling system with central heating and cooling coils. Each induction unit has a "secondary coil" that is capable of adding additional heating or cooling to the secondary air before mixing it with the primary air. The dissimilarities between different induction system designs are in the method of controlling the temperature of the supply air discharged into the conditioned space.

An induction unit can have a coil-control or a mixed-air control as the secondary air passes through the induction unit, as shown in Figure 18.2. With the coil control, the coil can be supplied with chilled or hot water. A modulating control valve responding to the demand of the room thermostat controls the amount of water flow through the secondary coil. A psychrometric plot of a coil-control induction unit is shown in Figure 18.3.

The secondary coil can be supplied by a four-pipe system or a two-pipe system. With the four-pipe system, the coil generally has two separate circuits, one for chilled water and one for hot water. Cooling and heating can be supplied to the induced air at any time. A four-pipe system also can have a single coil with changeover control such that the coil can use chilled water or hot water anytime as the load shifts between cooling and heating. With the two-pipe system, the secondary coil is supplied with chilled water in the summer and hot water in the winter. There are control limitations associated with the two-pipe system, and it also may require more primary air to achieve the required control range.

With the mixed-air concept, sometimes called face-and-bypass control, the secondary air is heated or cooled depending on the tempera-

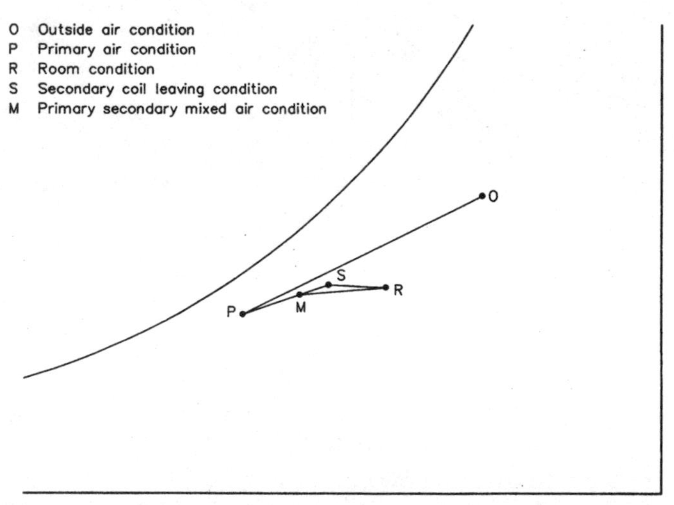

Figure 18.3. Psychrometric process of an induction unit with coil control.

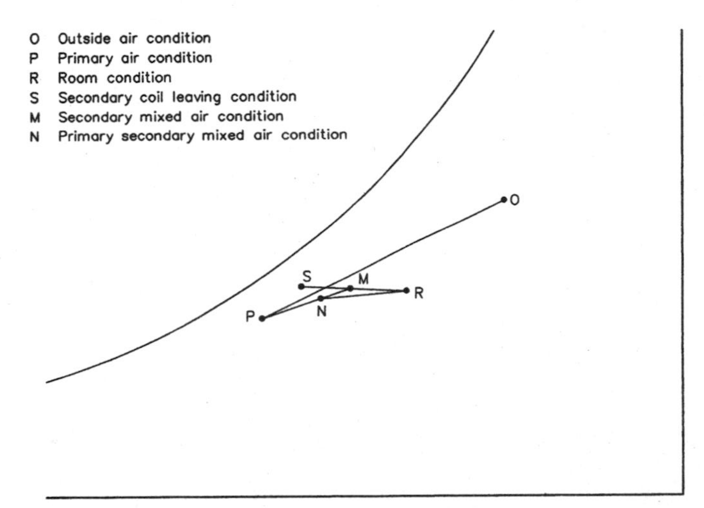

Figure 18.4. Psychrometric process of an induction unit with face-and-bypass control.

ture of the water flowing through the secondary coil, with the amount of secondary air flow through the coil controlled by a modulating damper that responds to the demand of the room thermostat. Under full load conditions, 100 percent of induced air flows through the secondary coil. As the load is reduced, part of the induced air will bypass the coil. Under minimum load conditions, 100 percent of induced air bypasses the secondary coil. A psychrometric plot of a mixed-air induction unit is shown in Figure 18.4.

PART 3

Variable Volume Systems

19
Variable Volume Systems with Perimeter Heating

The variable volume system concept was derived from the realization that the majority of air-conditioning loads in many buildings are cooling loads, that an air-conditioning system wastes energy primarily during reheat and mixing processes for temperature control, and that a "pure" variable volume system can be used successfully for any space requiring cooling only. These facts led engineers to analyze space load components of a typical building. People, lights, appliances, and solar loads are all cooling loads regardless of outside air temperatures. The only load component that requires both cooling and heating is the transmission load through the building skin. Since the transmission heat gain also can be grouped together with other cooling loads, the air-conditioning system can be greatly simplified and energy waste drastically reduced by using a variable volume cooling-only system throughout the building and providing a separate perimeter heating system to cope with periodic heat loss through the building skin. A schematic diagram of a typical variable volume system with perimeter heating is shown in Figure 19.1.

The main air handling unit for the variable volume cooling-only system is basically the same as that for the constant volume terminal reheat system, which consists of filters, cooling coils, and a supply fan. The only difference for this system is that the supply fan will have variable volume control. Since the unit is for cooling only, an economizer cycle should be used wherever possible to reduce cooling energy consumption. A preheat coil may be required if the system requires large amounts of outside air with low winter design temperature or if the system requires 100 percent outside air.

The main supply air duct extends from the air handling unit to branch ducts serving each functional space requiring independent temperature control. At each branch, a variable volume cooling-only supply air terminal will be installed and controlled by a room thermostat. Since the room temperature is controlled by varying the amount of air supplied into the room and there is no heating load for this system, there is no need for a heat coil to reheat or a hot air source for mixing. With this system, the amount of supply air varies in direct proportion to the room load. When the cooling load reduces to zero, the variable volume terminal shuts off completely.

Two types of terminals are used in variable volume systems. One is referred to as a *pressure-dependent terminal,* and the other, as a *pressure-independent terminal.* The pressure-dependent terminal is a very simple design consisting of a volume control damper that responds directly to the room thermostat. The thermostat positions the control damper to satisfy the space load. As terminals in a variable volume system change their supply air quantities, the static pressure in the main supply air duct changes. Since the volume damper is positioned by the thermostat, which does not respond to the pressure changes in the duct system, once the pressure in the duct changes, the amount of air supplied through the damper will increase or decrease depending on the system pressure variation. As more or less air supplied to the room changes the room temperature as a result of pressure variations, the room thermostat will sense the temperature change and reposition the damper accordingly to maintain the room temperature.

The pressure-independent terminal has an added differential pressure sensor or velocity pressure sensor in the control loop. The room thermostat, instead of controlling the position of the variable volume damper directly, sends a signal to a reset controller that controls the amount of supply air required to satisfy the space load. When the

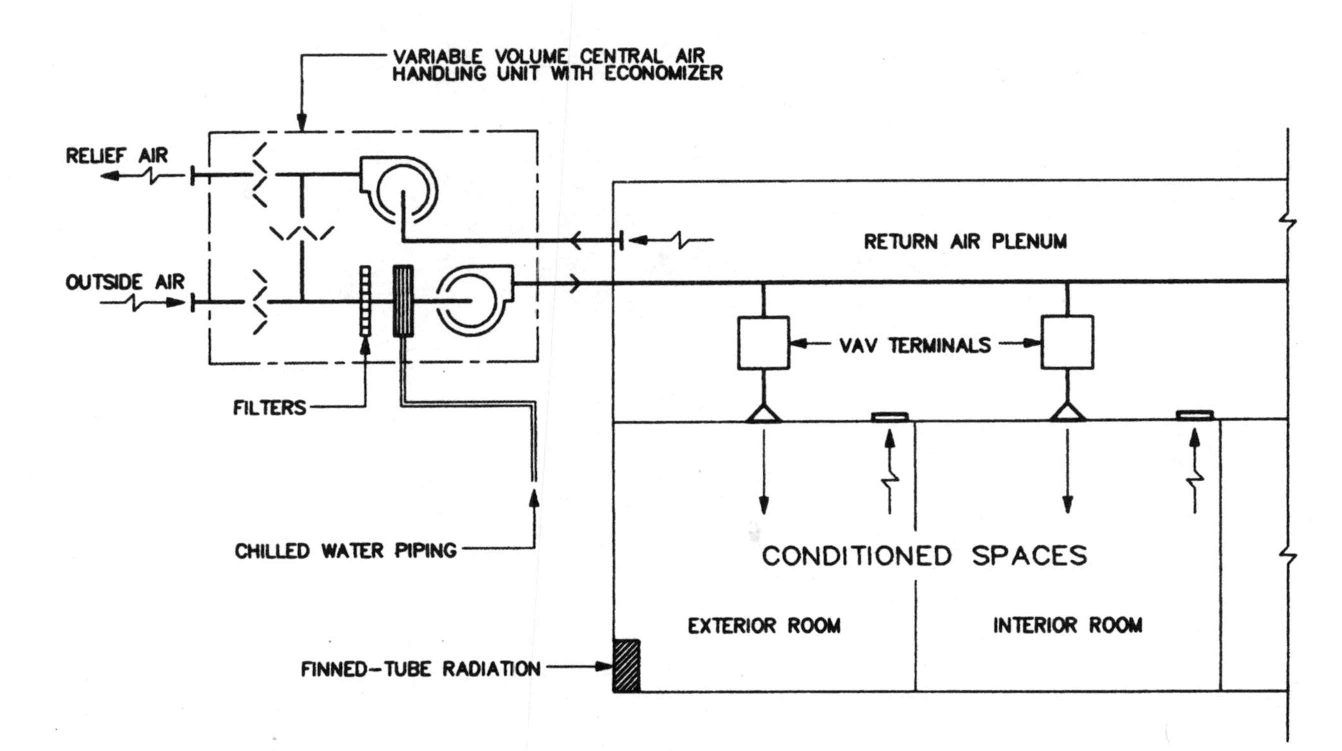

Figure 19.1. Variable volume system with perimeter heating.

velocity sensor senses a change in duct velocity due to upstream pressure variations, it sends a reset signal to the controller to modulate the volume damper position to maintain the required duct velocity and thus the required supply air quantity. With this control, the room thermostat is isolated from the influence of pressure changes in the duct system, and the space temperature will experience fewer fluctuations.

There should be no minimum setting for the cooling-only variable volume terminal. When there is no load in the space, a minimum amount of supply air, however small, can overcool the conditioned space over a period of time, and it may take a relatively long time to warm the space up to a temperature suitable for occupancy. For this reason, a pure variable volume system sometimes is viewed as a less flexible system. For example, this system cannot satisfy the varying cooling load needs of an area in a building that requires a fixed amount or a minimum amount of air circulation. It will be considerably more difficult also for a pure variable volume system to satisfy the condition where pressure balance between spaces is a requirement.

Sometimes, designers who are more familiar with the constant volume concept or whose concern stops at the peak design load only may develop a false sense of security in designing a variable volume system. After having sized all variable volume terminals for peak design air quantity, they may erroneously feel that they have provided adequate ventilation for each conditioned space. One must remember that regardless of the size of the supply air terminal, the amount of supply air in a variable volume system will track the actual load at any given time and the terminal will continuously vary the supply air volume to follow the load. When the load diminishes to zero, there will be no air supplied into the conditioned space and thus no ventilation.

There are concerns that in order to provide minimum ventilation at all times, variable air supply to a conditioned space should never be allowed to reduce to shutoff. This concern can put a serious limitation on application of a pure variable volume system. One should remember, however, that a pure variable volume system is built on the concept that when there is no load in the space, the space is unoccupied. To shut off the supply air under this condition is the most effective way to conserve energy, much the same as fan coil unit applications, where occupants are asked to turn off the unit when leaving the room, or applications where air-conditioning systems are shut off on weekends for energy conservation. With this in mind, the concern should be modified to consider that in order to provide effective minimum ventilation at all times, air supply to a conditioned space should never be allowed to shut off at any time *unless the space is unoccupied.*

Cooling load through fenestration consists of both transmission and radiant components. Design engineers should be cautioned that resultant values for transmission and radiant loads through fenestration are combined in many load calculation procedures. The peak cooling load for a space with southern exposure can, however, occur in winter when outside air temperature is lower than the indoor design temperature. In this case, the transmission loss neutralizes a portion of the radiant heat gain, but the radiant load component in winter alone will be higher than the sum of transmission and radiant loads for all other hours of the year. Figure 19.2 shows an example of an hour-by-hour load calculation output that illustrates this effect. Under these conditions, the conventional cooling load calculation procedure should be modified to delete the transmission loss credit, which, with this particular design concept, has been taken care of by a separate heating system, and the cooling-only supply air for these south-facing spaces will have to be increased to eliminate the heat loss credit to the cooling load.

At the air handling unit, care must be taken in determining the mixed air temperature for a variable air volume system using a fixed minimum amount of outside air. The purpose of calculating this mixed air temperature is to determine the need and/or amount of preheat. Since preheat will only be needed during the heating season, most likely the variable volume system will not be operating at full cooling capacity. Consequently, the total supply air quantity, and in turn the return air quantity, under partial load conditions will be lower than the design air quantity. Thus the mixed air temperature of a fixed amount of outside air and a reduced amount of return air can be considerably lower than that if the maximum air quantity is used (often it can be lower than the required supply air temperature).

It is important to maintain the minimum required amount of outside air at all times to satisfy the ventilation requirement, even as the volume control varies the total supply quantity. Special attention should be focused on providing reliable control schemes to achieve this goal. Where an economizer cycle is used in conjunction with this system, the economizer control, while maintaining the mixed air temperature to satisfy the cooling needs, also must maintain not less than the required minimum amount of outside air throughout the variable volume range of the system. It should be noted that if the system is serv-

AUTOMATED PROCEDURE FOR ENGINEERING CONSULTANTS, INC. HCCL-I
SAMPLE PROBLEM - 8 STORY OFFICE BUILDING HA 6094 T. Y. SUN 6/20/82

ROOM 301 PEAK AND HOURLY LOADS PEAK AT 2 PM, DEC

ROOM NO.	PEAK MO/HR	SENS. C. LOAD	TOTAL C. LOAD	SENS. RATIO	ROOM CFM	HEATING LOAD	ROOM AREA	CFM /SF	NO. PEOP.	DT HTG.
301	12/14	76043.	79643.	.95	3471.	13281.	1800.	1.93	18	3.5

HOUR	8	9	10	11	12	13	14
SENSIBLE							
WINDOW TRANS	-7260.	-4937.	-3775.	-2033.	-871.	-290.	-290.
WINDOW SOLAR	7040.	18778.	31679.	42894.	50865.	55116.	55018.
WALL TRANS	-1798.	-1222.	-935.	-503.	-216.	-72.	-72.
WALL SOLAR	1538.	963.	983.	1115.	1407.	1812.	2200.
ROOF TRANS	0.	0.	0.	0.	0.	0.	0.
ROOF SOLAR	0.	0.	0.	0.	0.	0.	0.
FLOOR	0.	0.	0.	0.	0.	0.	0.
PARTITION	0.	0.	0.	0.	0.	0.	0.
INFILTRATION	0.	0.	0.	0.	0.	0.	0.
LIGHT TO RM	10771.	13219.	14688.	14688.	14688.	14688.	14688.
PEOPLE	2200.	3500.	4300.	4500.	4500.	4500.	4500.
APPLIANCE	0.	0.	0.	0.	0.	0.	0.
TOTAL SEN.	12491.	30301.	46940.	60661.	70373.	75754.	76043.
LATENT							
INFILTRATION	0.	0.	0.	0.	0.	0.	0.
PEOPLE	2880.	3600.	3600.	3600.	3600.	3600.	3600.
APPLIANCE	0.	0.	0.	0.	0.	0.	0.
TOTAL	2880.	3600.	3600.	3600.	3600.	3600.	3600.
TOTAL LOAD	15371.	33901.	50540.	64261.	73973.	79354.	79643.

HOUR	15	16	17	18	0	HEATING LOAD FOR 34. F OSA	
SENSIBLE							
WINDOW TRANS	-290.	-871.	-2614.	-3775.	0.	WINDOW	10692.
WINDOW SOLAR	49480.	35065.	13770.	4965.	0.	WALL	2589.
WALL TRANS	-72.	-216.	-647.	-935.	0.	ROOF	0.
WALL SOLAR	2425.	2591.	2830.	2560.	0.	FLOOR	0.
ROOF TRANS	0.	0.	0.	0.	0.	PART.	0.
ROOF SOLAR	0.	0.	0.	0.	0.	INFILT.	0.
FLOOR	0.	0.	0.	0.	0.	TOTAL	13281.
PARTITION	0.	0.	0.	0.	0.		
INFILTRATION	0.	0.	0.	0.	0.	SUMMARIES	
LIGHT TO RM	14688.	14688.	14688.	14688.	0.		
PEOPLE	4500.	4500.	4000.	2300.	0.	AREA	1800.SF
APPLIANCE	0.	0.	0.	0.	0.	VOLUME	16200.CF
TOTAL SEN.	70731.	55757.	32027.	19803.	0.	PEOPLE	18
						INFILT.	0.CFM
LATENT						BTU/SF CLG.	44.2
INFILTRATION	0.	0.	0.	0.	0.	BTU/SF HTG.	7.4
PEOPLE	3600.	3600.	2880.	720.	0.		
APPLIANCE	0.	0.	0.	0.	0.	SUPPLY AIR AT 20	
TOTAL	3600.	3600.	2880.	720.	0.	F DT	3471.CFM
						AC/HR.	12.9
TOTAL LOAD	74331.	59357.	34907.	20523.	0.	HTG DT =	3.5 F

Figure 19.2. Example of an hour-by-hour load calculation output.

ing spaces requiring humidity control, the cost savings for using less cooling energy with an economizer cycle should be compared with the additional cost of humidifying larger amounts of outside air.

A variable volume system does not have the excellent humidity control characteristic of a constant volume terminal reheat system, but it offers better humidity control than the conventional double-duct system. Figure 19.3 shows the psychrometric process of a cooling-only variable volume system. Comparing Figure 19.3 with Figure 8.2 for the terminal reheat system and Figure 9.3 for the double-duct system clearly illustrates the differences.

The total quantity of supply air in any of the constant volume system discussed in previous chapters is the sum of the quantity of supply air required to satisfy the instantaneous peaks of each conditioned space. In a variable volume system, however, since the quantity of supply air for any space under partial load conditions will be reduced proportionately according to the load requirements, the total quantity of supply air can be sized for the block load of the building, which can be significantly less than that required for constant volume systems. Consequently, the size of the supply fan and the motor horsepower for a variable volume system will be correspondingly smaller.

Volume control should be installed at the supply fan to prevent

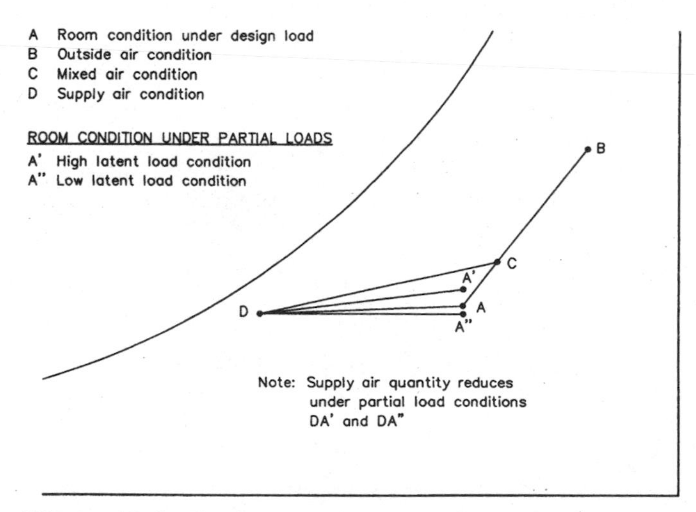

Figure 19.3. Psychrometric process of a cooling-only variable volume system.

excessive pressure buildup in the duct system as variable volume terminals reduce supply air quantities under partial load conditions. Depending on the type of volume control device chosen, transportation energy can be further reduced under partial load conditions that exist at almost all times in an air-conditioning system. The amount of transportation energy savings varies from practically none with bypass control to maximum for variable-speed drive. Discharge dampers, scroll volume control for forward-curved fans, variable-pitch control for vaneaxial fans, variable-inlet vanes, variable-wheel-width controls for centrifugal fans, and variable-frequency drives to change fan speed are all viable methods for volume control. Detailed discussions on the pros and cons of these volume control devices can be found in the *ASHRAE Handbook* series and many other related textbooks.

Energy savings, however, should not be the only consideration in choosing what type of volume control to use for a given project. The consideration should include such factors as the size of the system, first-cost penalty, reliability of the control device, ease of maintenance and repair, and acoustical and electrical noise generation, to name a few.

There are several methods to provide heating to the building perimeter. The simplest way is to use finned-tube radiation or convectors along the perimeter of the building, as shown in Figure 19.1. Hot water, steam, or electricity can be used as the energy source. Another way of providing perimeter heating is to use radiant panels along the building perimeter. Most of these panels are electric-powered, with occasional hot-water applications. Radiant panels can be either ceiling- or wall-mounted. Since the finned-tube radiation takes a certain amount of floor space along the building perimeter, the ceiling-mounted radiant panel application may have better space utilization. On the other hand, finned-tube radiation is considerably more effective to offset the cold down drafts near the window.

The ideal way to control these types of perimeter heating systems is by sensing the inside surface temperature of the fenestration. Sometimes additional thermostats located on the inside surface of the exterior wall can be used to control the heating system.

The transmission heat loss also can be neutralized by a single-zone constant volume heating-only air handling system. Supply air quantity for each segment of the building perimeter is calculated to cope with the maximum transmission loss for that segment. Since the transmission load varies in direct proportion to the outside air temperature, a simple reset schedule can be used to control the supply air temperature, which varies in inverse proportion to the outside air temperature. No local control in the conditioned space is required. When the outside air temperature approaches the room temperature, the heating system can be shut off.

Since the sole purpose of this heating-only air handling system is to neutralize the transmission load, and since the supply air temperature of this system is always inversely proportional to the outside air temperature, introducing outside air to this system is unnecessary, and the system should always use 100 percent return air without an economizer cycle. A schematic diagram of this system using rooftop units is shown in Figure 19.4. Obviously, the evaporator for the variable volume system and the gas furnace for the perimeter system can be a chilled-water coil and a hot-water coil, respectively.

The use of a centrally located constant volume heating system for the building perimeter in conjunction with a variable volume cooling-only system is often referred to as a *dual-conduit system.* A broader scope version of the dual-conduit system is discussed in the next chapter.

Supply air from this constant volume heating system is distributed through a series of registers or slot diffusers located along the perimeter of the building, as shown in Figure 19.4. The amount of air supplied through each diffuser is sized to match the transmission loss for the section of the building perimeter the diffuser covers. Since the facade of the building is rarely modified, manual air balance using balancing dampers is generally satisfactory. Constant volume terminals are sometimes used for the perimeter system that will respond to duct pressure variations and maintain constant supply air volume at all time.

The drawback of having a set of perimeter supply outlets is that, for the exterior spaces, there will be two sets of supply outlets for each conditioned space. A design variation on the supply outlets for the perimeter system is to introduce the perimeter supply air downstream of the variable volume terminal serving exterior spaces so that only one set of diffusers will be required for each conditioned space.

The heating air handling unit is generally a small unit compared with the central cooling-only variable volume unit. Since the skin system uses 100 percent return air, and considering that a return air plenum is commonly used for a general office building, it is possible to use a series of heating-only fan coil units locally installed in the ceiling return air plenum to neutralize the heat loss through the building skin and thus eliminate the need for a centralized heating system. Figure 19.5 shows a schematic diagram of this arrangement together with a combined supply diffuser, as discussed in the preceding paragraph.

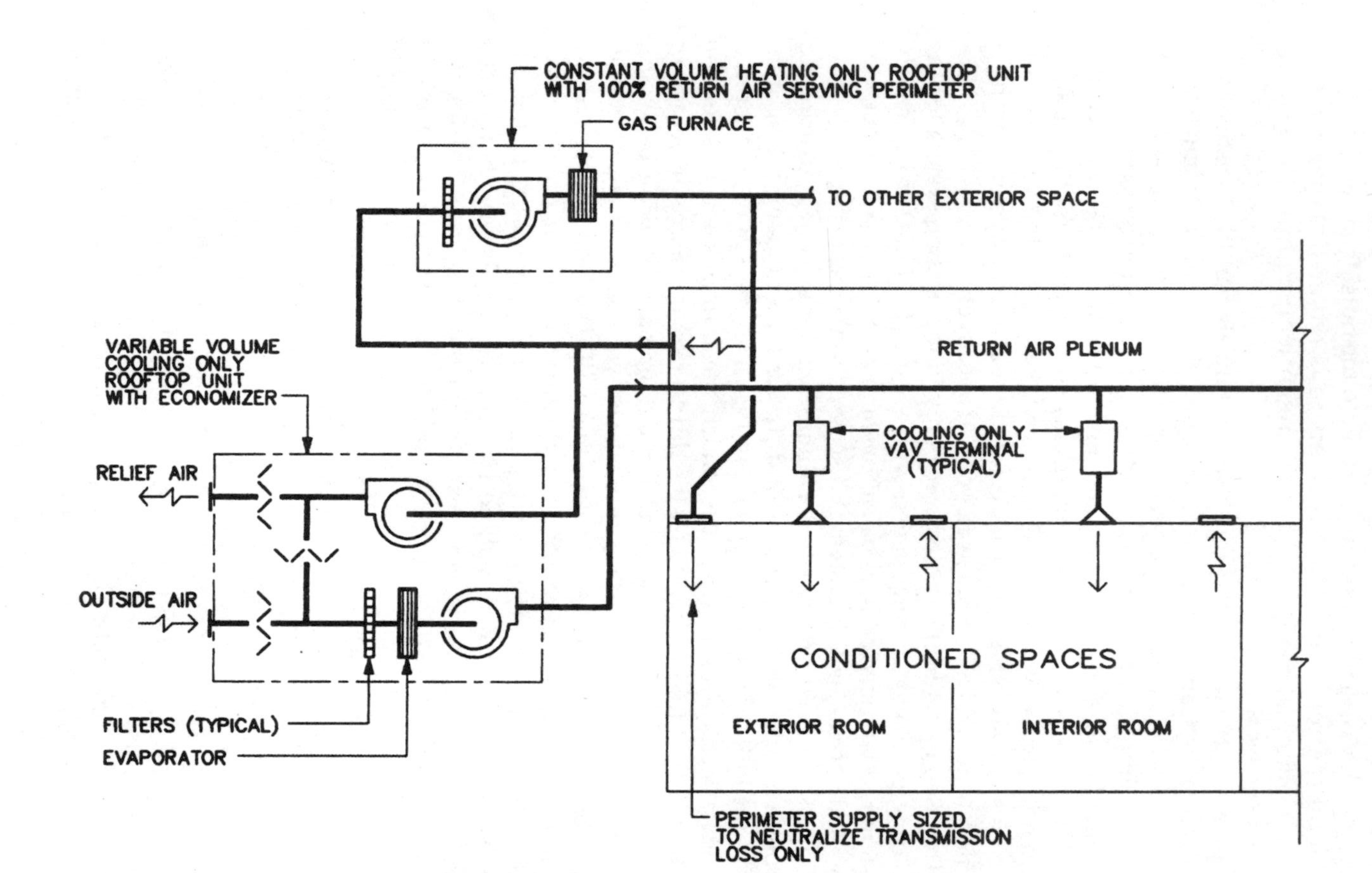

Figure 19.4. Variable volume system with heating-only perimeter system.

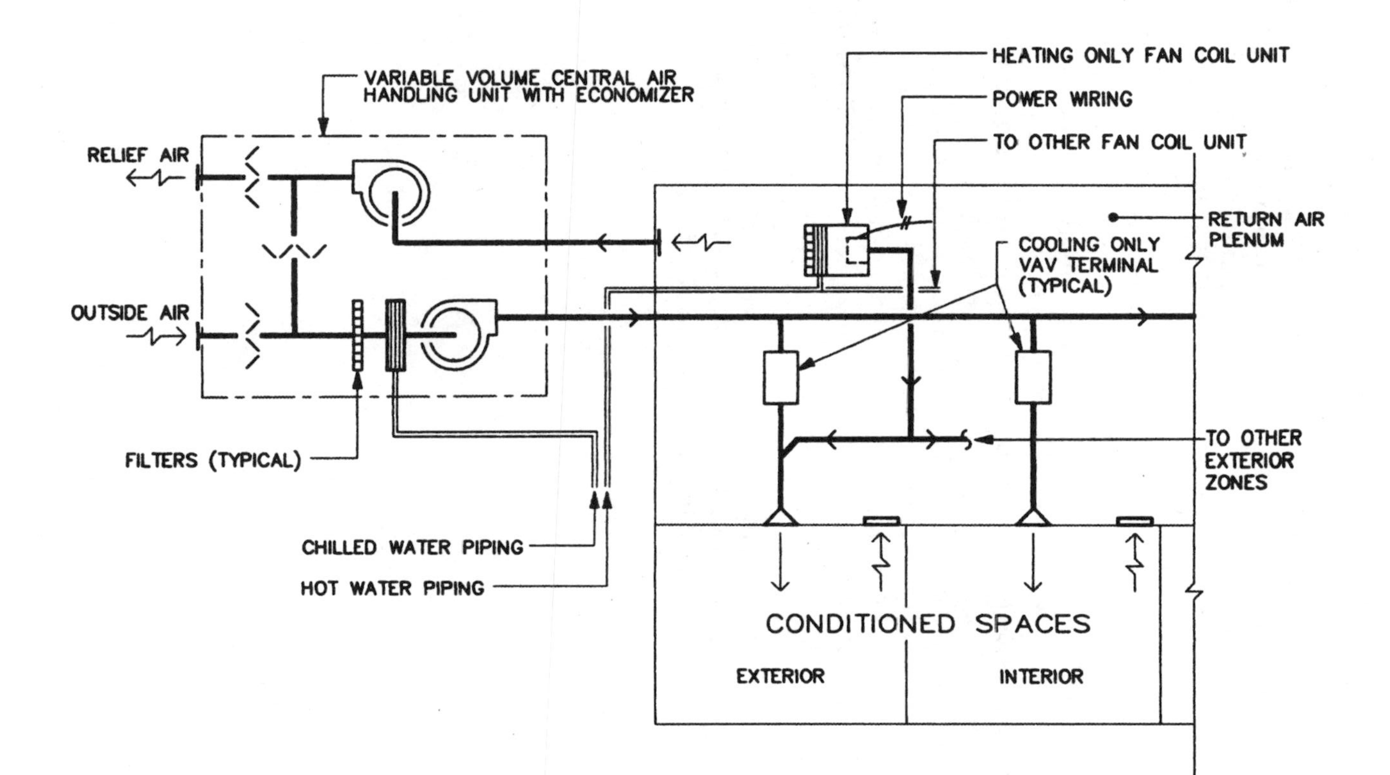

Figure 19.5. Variable volume system with fan coil units serving building perimeter.

In designing the heating system, one must make sure that all areas that may have a potential transmission loss are covered by the system. Such features as the floor of an underground parking garage, the basement wall and slab, and the second floor slab over a soffit above the first floor can have enough heat loss to cause insufficient heating by a conventional system. It is important to note that if the transmission loss is not neutralized by the skin system, it may cancel out the internal heat gain and cause a variable volume terminal to shut off even when a space is occupied. This will deprive occupants of the required minimum ventilation and cause indoor air quality (IAQ) concerns. For this reason, a pure variable volume system may not be the ideal choice for serving a basement, particularly if the basement has a very low design lighting load.

The increasing concern for IAQ issues may have placed some restrictions on the applications of this system. Judging from many existing and successful applications of this system, however, the design concept is sound. With proper maintenance, careful control strategy designed to ensure minimum required outside air supply to the system at all times, and the building industry actively seeking new ways to control indoor air quality by improving materials used in the building interior, there should be no difficulty with use of this energy-conserving system for most general building air-conditioning applications.

20
Dual-Conduit Systems

This system is a modification and expansion of the variable volume system with a constant volume heating air handling unit discussed in the preceding chapter. Since both transmission heat loss and transmission heat gain through the building skin are continuous linear functions of the outside air temperature, by adding a cooling source in the perimeter heating unit (see Fig. 19.4), the same unit and its duct distribution system can take care of transmission heat gains in the cooling season.

With this system, the size of the variable volume component will be reduced slightly because the cooling load for transmission heat gain is shifted from the variable volume system to the constant volume perimeter system. The distribution ductwork for the constant volume perimeter system will remain practically the same, since the same distribution system will deliver hot air in winter and cold air in summer. For the same reason as stated in Chapter 19 for the heating-only perimeter air handling system, the constant volume perimeter system should use 100 percent return air without an economizer cycle. A schematic diagram of a typical dual-conduit system is shown in Figure 20.1. The only difference between Figures 19.4 and 20.1 is the added cooling source in the perimeter air handling unit in Figure 20.1 for the extended coverage of transmission heat gains into the cooling season.

With this perimeter air handling system, the quantity of supply air for each segment of the building perimeter will be designed to cope with the maximum transmission heat gain and maximum transmission heat loss. The air quantity is generally determined for the maximum cooling load based on a predetermined cold supply air temperature. The hot air temperature is derived from the maximum heating load with the calculated quantity of supply air using Eq. (9.1). Since the

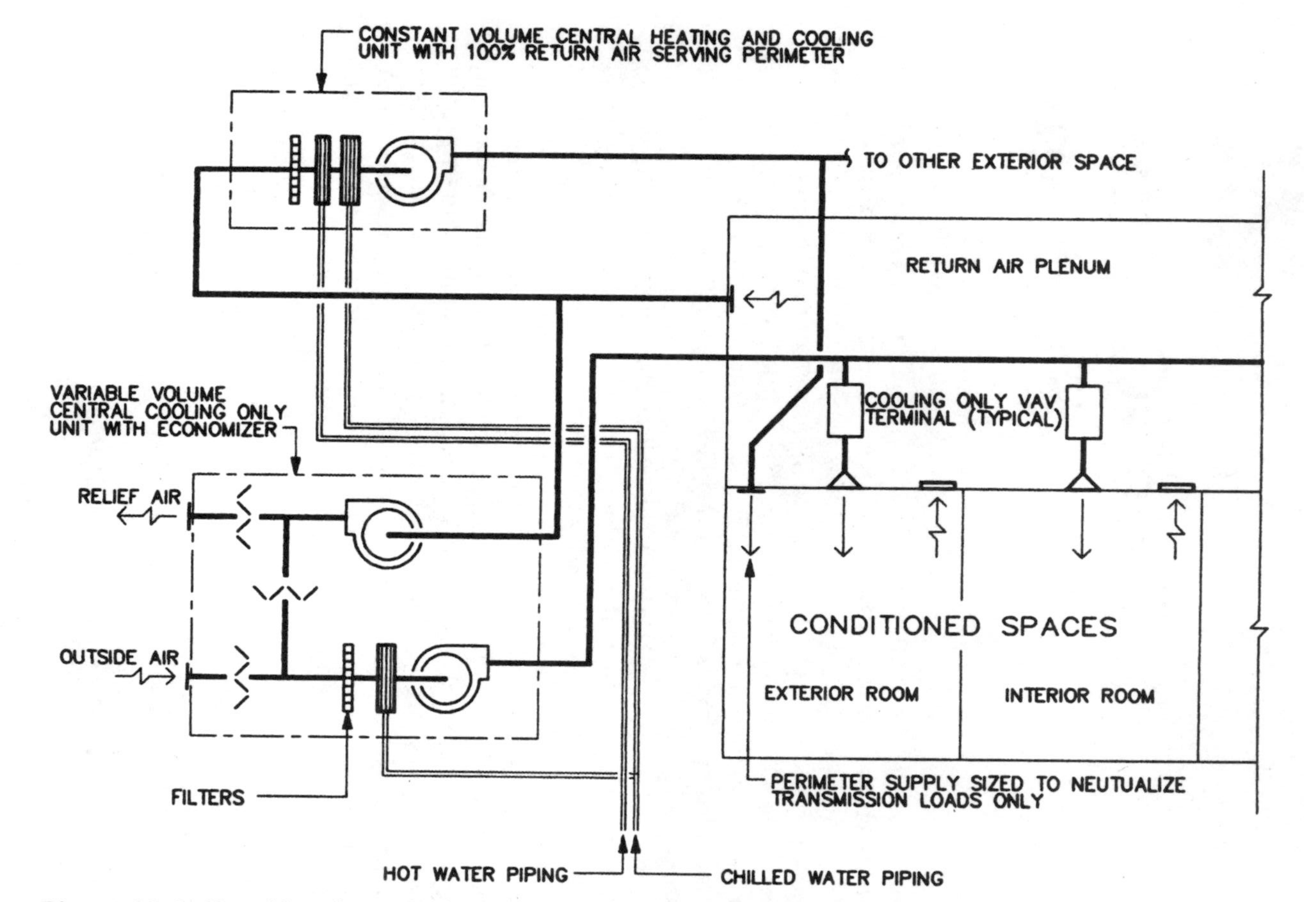

Figure 20.1. Variable volume dual-conduit system. *Note:* Cooling coil may be omitted if the variable volume cooling unit and supply duct system are sized for transmission gain also.

variation in transmission load is linear with respect to variations in outside air temperature, a simple reset schedule can be used to control the supply air temperature throughout the heating and cooling range. In other words, supply air temperature is controlled between the maximum design temperature in the winter and the minimum design temperature in the summer and varies in inverse proportion to the outside air temperature.

The variable volume cooling-only system for this dual-conduit system is identical to the variable volume system with perimeter heating discussed in Chapter 19. With the dual-conduit system, since the transmission load is neutralized by a separate air handling system, the cooling load used for calculating peak supply air quantity for exterior spaces should exclude the transmission loads through the building perimeter. This will reduce the peak supply air quantities of the variable volume cooling-only system for most of the exterior spaces. Again, one must be cautioned that for exterior spaces where peak cooling load occurs in winter, when the outside air temperature is lower than the space temperature, transmission loss must not be credited to the cooling load. See the discussion on this topic in Chapter 19.

Note that instead of using an evaporator and a gas furnace for the system, as shown in Figure 19.4, a chilled-water coil is used for the variable volume system and chilled- and hot-water coils are used for the perimeter unit in Figure 20.1. Use of this dual-conduit system is efficient and more convenient where the air handling systems are served with chilled and hot water from a central plant. If the application requires the use of self-contained rooftop units, it may be more reasonable to use a heating-only unit to take care of transmission heat loss and let the central variable volume cooling unit take care of the transmission heat gain, as shown in Figure 19.4. Otherwise, a complete condensing unit will have to be added to the perimeter unit for the additional cooling duty, which adds more cost to the system.

21 Single-Fan Dual-Conduit Systems

The single-fan dual-conduit system is a design variation of the conventional dual-conduit system that uses two air handling units: a constant volume, variable temperature air handling system for the building transmission load and a constant temperature, variable volume air handling system for all other loads (see Fig. 20.1). This system uses a single air handling unit to perform both duties.

The air handling unit configuration of this system is very similar to that of a double-duct system. The difference is that with this system, a cooling coil is added in series with the heating coil in the hot plenum. The supply air duct extended from this heating/cooling plenum is dedicated to take care of the transmission heat gain and heat loss throughout the building perimeter served by the system. The cooling coil in the cold air plenum is identical to that for the conventional double-duct system. The cold air duct that extends from this plenum performs the same duty as the constant temperature, variable volume system of the conventional dual-conduit system that serves all other cooling loads. Being a variable volume system, the supply fan for this system should have variable volume control. A schematic diagram of a typical single-fan dual-conduit system is shown in Figure 21.1.

The main cooling coil and the variable volume cooling-only supply air duct of this system also can be sized to take care of the transmission heat gain through the building perimeter. Under such a condition, the cooling coil serving the building perimeter can be deleted, as shown in Figure 21.2. This system will perform much like a variable volume dual-conduit system with a heating-only perimeter air han-

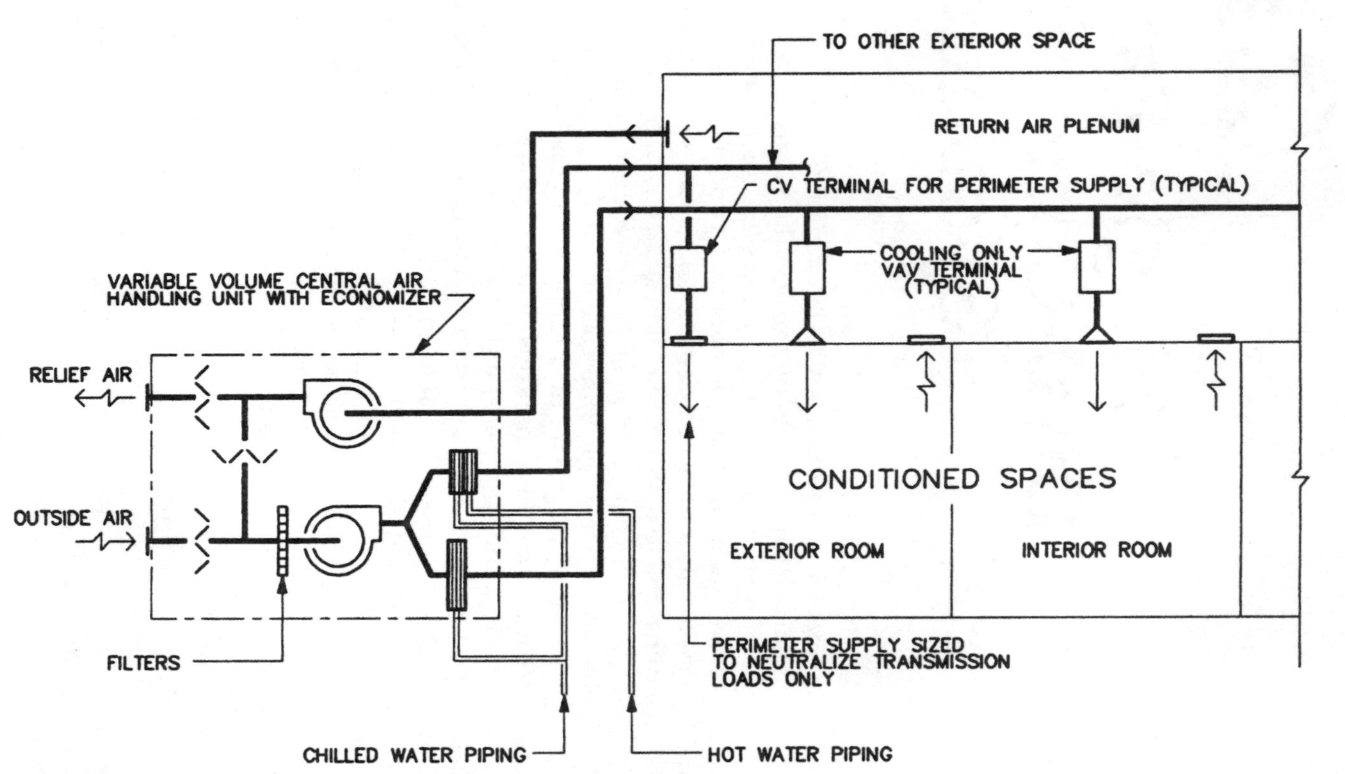

Figure 21.1. Single-fan variable volume dual-conduit system.

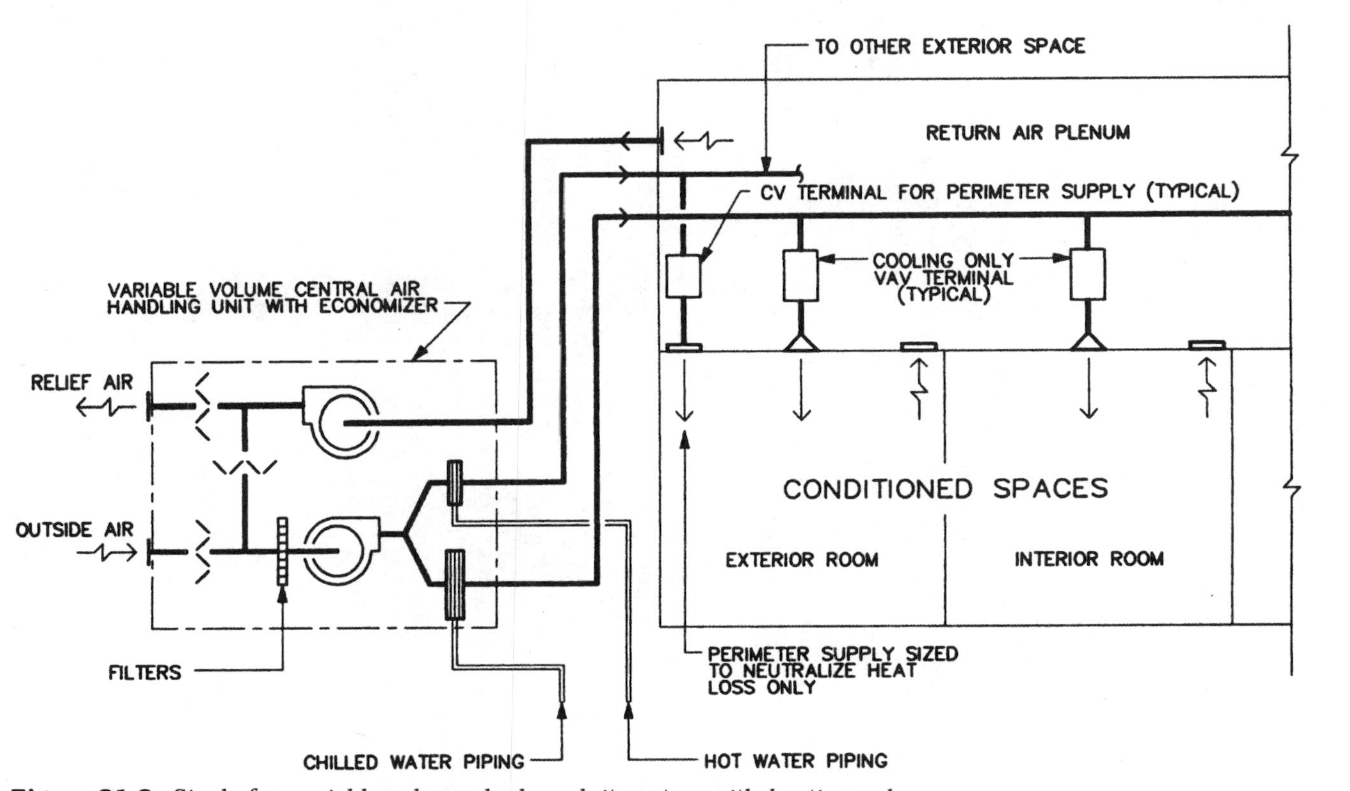

Figure 21.2. Single-fan variable volume dual-conduit system with heating only for perimeter.

dling unit, as shown in Figure 19.4. It is interesting to note that in this case, the components in the air handling unit will be the same as those of the constant volume double-duct system discussed in Chapter 10 and the variable volume double-duct system discussed later in Chapter 25. It must be noted, however, that the design concept and system operation of a single-fan dual-conduit system are distinctly different from those of the various forms of double-duct systems discussed in these related double-duct chapters.

The air distribution system for the single-fan dual-conduit system is similar to that for the conventional dual-conduit system. The only difference is that with this system, constant volume terminals must be used for the perimeter supply outlets so that proper supply air volume can be maintained for all perimeter outlets. With the conventional two-fan dual-conduit system, since there is a separate air handling unit dedicated to constant volume supply to all perimeter outlets, the use of constant volume terminal becomes only a desirable option.

With only one air handling unit, the cost of this system should be less than that of the two-fan conventional dual-conduit system. An economizer cycle may be used for this system to conserve energy. However, since the supply fan supplies air to both the cooling-only variable volume system and the perimeter system, the economizer cycle will supply cold air to both systems when the outside air temperature is below the supply air temperature. Compared with the conventional dual-conduit system, which uses 100 percent return air for the perimeter system, there is an energy waste associated with this arrangement during heating season.

This system has another potential operational drawback. The cooling process in an air-conditioning system, in addition to lowering the supply air temperature, also condenses and removes excess moisture to dehumidify the conditioned air and maintain a reasonable humidity level in the conditioned space. The source of the excess moisture is primarily the outside air when it has a high moisture content. The design concept of a dual-conduit system calls for the cooling coil in the perimeter system to be deactivated when the outside air temperature drops below the space temperature. During this period, therefore, the perimeter system does not have the dehumidification capability. During rainy weather or mild weather conditions when outside air is humid, any amount of humid outside air introduced into the perimeter system will come through the system and go directly into the conditioned space, raising the humidity level inside. The problem is more pronounced where an economizer cycle is incorporated in the system. During mild weather, a larger amount of outside air can be introduced by the economizer cycle through the perimeter system into the conditioned space without dehumidification. For this reason, the conventional two-fan dual-conduit system with the perimeter system using 100 percent return air is generally a better choice for dual-conduit system applications.

22 Cooling-Only Variable Volume Systems with Constant Volume Air Handling Units

A cooling-only variable volume system as described in previous chapters will vary the volume of supply air at each terminal serving the conditioned space as the cooling load varies in the conditioned space. At the same time, the supply air volume of the air handling unit should vary accordingly to prevent over-pressurization of the supply ductwork. To achieve this, additional devices and controls will be required at the supply fan in a variable volume air handling unit.

The cooling-only variable volume system also can be designed using a constant volume supply fan in the air handling unit in conjunction with variable volume terminals serving various conditioned spaces. There are two application variations of this design concept. One method is to add a bypass duct at the air handling unit, and the other is to use bypass terminals for each thermostatically controlled space.

Constant Volume Air Handling Unit with Bypass Duct

Adding a bypass duct at the air handling unit for the variable volume system will allow the system to use a conventional constant volume air handling unit. A schematic diagram of this system is shown in Figure 22.1. As the variable volume terminals reduce supply air volume, a pressure sensor in the main supply duct senses the pressure increase in the duct and modulates open the normally closed damper in the bypass duct to allow a portion of the supply air to return directly to the return side of the air handling unit.

With the added bypass duct, the supply fan in the air handling unit will not notice the volume change downstream in the air distribution system. Besides the control damper in the bypass duct, no volume control device is required for the supply fan. The bypass of supply air will prevent excessive pressure buildup in the supply duct system, which will make the system operate more quietly. The operation of this type of system is very simple.

The drawback of this system is that there will be no transportation energy savings because the air handling unit will operate at constant volume at all time. For small systems, however, the fan horsepower is relatively small, so it is difficult to justify the use of a true variable volume system. Under such circumstances, this type of system can be used very effectively.

Constant Volume Air Handling Unit with Bypass Terminals

The use of bypass terminals, sometimes referred to as the *spill terminals* or *spill boxes,* can be viewed as a decentralized bypass duct arrangement. It is a common practice for an air-conditioning system to use ceiling space as a return air plenum. The supply air terminal can be designed so that when the space does not require full cooling, a portion of the supply air can be diverted into the return air plenum without entering the conditioned space. The reduced amount of the supply air entering the conditioned space will satisfy the space needs. The portion of supply air "spilled" into the return air plenum will mix with return air from the conditioned space and flow back directly to the air handling unit. Since the air supplied to the bypass terminal

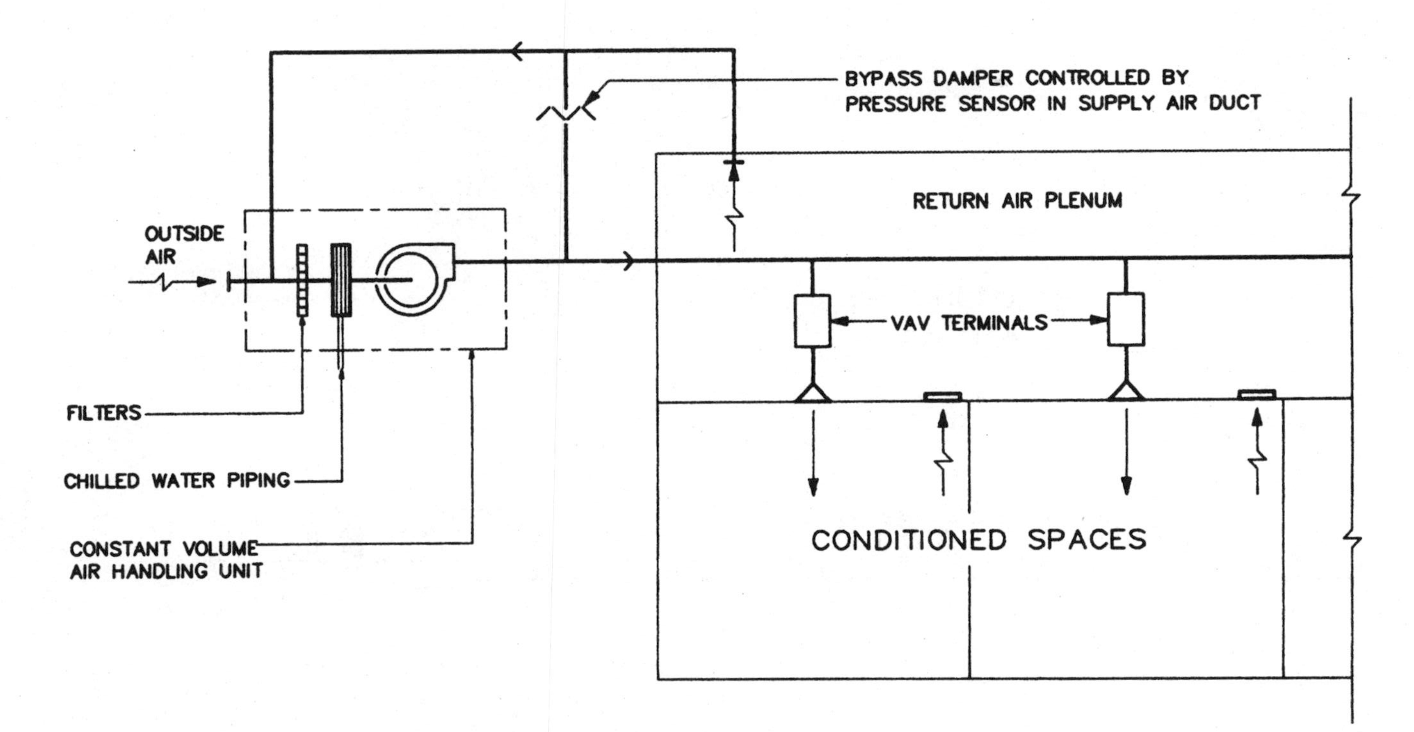

Figure 22.1. Variable volume system with a bypass duct at the air handling unit.

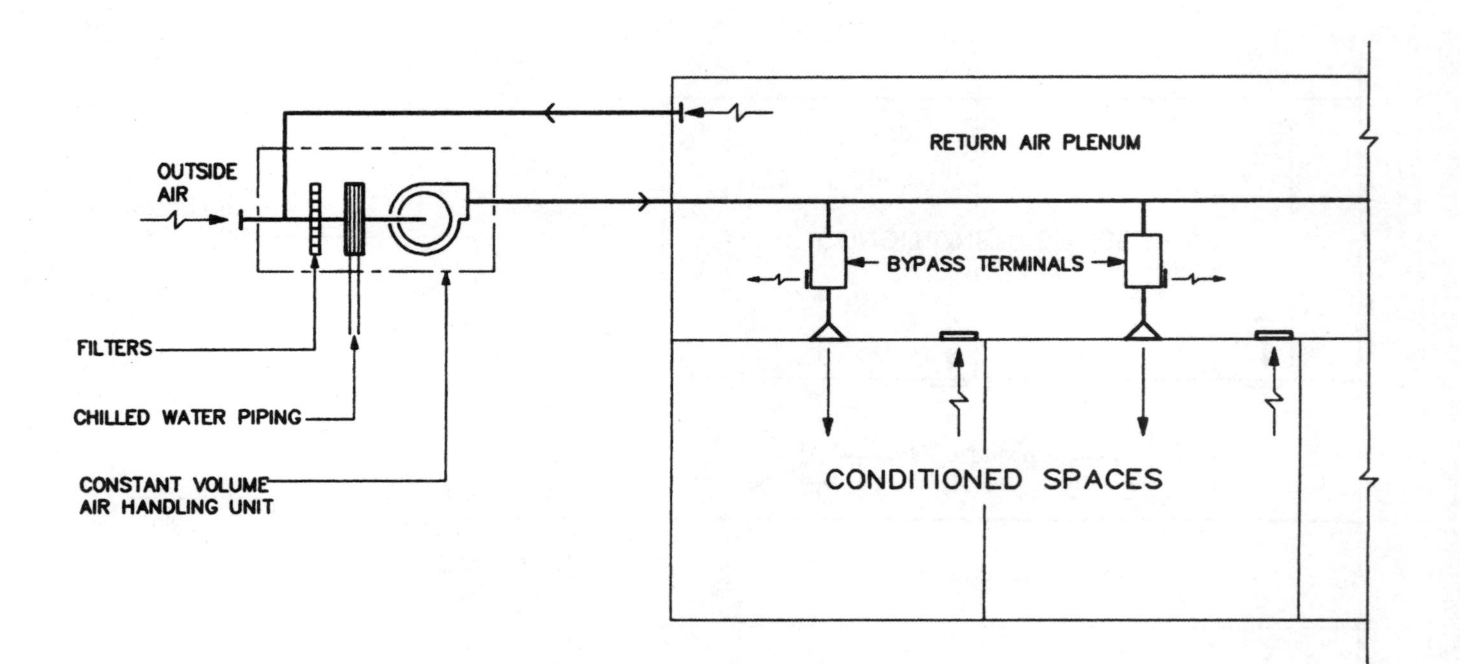

Figure 22.2. Variable volume system with bypass terminals.

does not change, the supply fan will not notice the reduction in supply air to the conditioned space. Thus no volume control device is required at the supply fan, and operation of the system is simplified. A schematic diagram of this type of system is shown in Figure 22.2.

Like the system with a bypass duct at the air handling unit, the bypass terminal system does not offer any transportation energy savings. The system is generally applicable only to small buildings, where a true variable volume system is difficult to justify. It should be cautioned that problems may arise when applying this system to spaces where loads vary significantly. The conditioned space may be overcooled, since under low load conditions, the bypass terminal will spill large amounts of supply air into the return air plenum directly above the conditioned space, turning the ceiling over the conditioned space into a radiant cooling panel that may subcool the conditioned space. For this reason, systems with bypass terminals are now used less frequently.

Coupled with any heating scheme to take care of perimeter heat loss, a variable volume system with a constant volume air handling unit can be used for any variable volume application where transportation energy costs using variable volume control at the supply fan are difficult to justify. This is particularly true with the bypass duct design option. When properly applied, this system can be very effective in thermal energy conservation for smaller system applications.

23 Variable Volume, Variable Temperature Systems

Variable volume, variable temperature systems are a further development of the variable volume systems using constant air handling units discussed in Chapter 22. As mentioned in Chapter 7, a typical variable volume system is a cooling-only system. A separate heating system or a separate means for heating at each thermostatically controlled terminal is needed for heating. The primary goal of the variable volume, variable temperature system discussed here is to create a system capable of performing both heating and cooling, thus eliminating the need for a separate heating system commonly associated with variable volume cooling-only systems. A schematic diagram of this type of system is shown in Figure 23.1.

The system was originally developed to solve the problem of using a constant volume single-zone heating and cooling air handling unit to serve multiple spaces with different load characteristics, as shown in Figure 2.1. On the cooling side, this system is identical to the variable volume system with a bypass duct discussed in Chapter 22. To satisfy the cooling load variations and provide individual comfort, variable air volume terminals are used in conjunction with a bypass duct at the air handling unit. To fine-tune control and conserve energy, an electronic control system is installed to monitor and control the variable

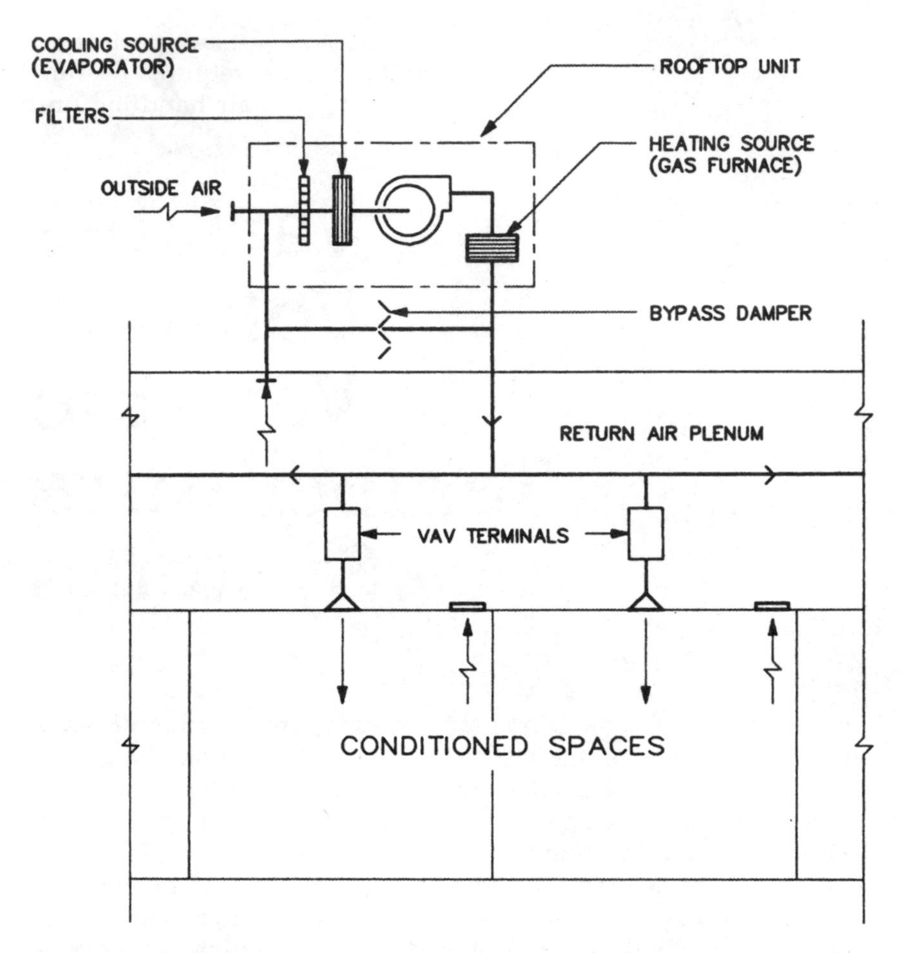

Figure 23.1. Variable volume, variable temperature system.

volume terminals. By scanning the cooling demands in each variable volume terminal, the temperature of the supply air is constantly reset to satisfy the terminal requiring maximum cooling. When the cooling demand of all terminals is satisfied, the temperature of the supply air will "creep up" so that at least one of the terminals is delivering 100 percent of its supply air volume. With a constant volume air handling unit, raising supply air temperature will reduce the load to the cooling coil.

A heating coil is also installed in the air handling unit of the variable volume, variable temperature system. As the electronic control scans the system, a terminal can be at its minimum volume setting while the room temperature continues to fall; i.e., the space requires heating. Under such a condition, the control switches the air handling unit from cooling to heating. As the terminal requiring heating senses the temperature of the supply air rising above the space temperature, the volume control at the terminal will open to supply the maximum amount of air to heat the space quickly. While the system is supplying heated air to satisfy one or a few terminals requiring heating, hot air also will be supplied to the rest of the terminals, which are under varying cooling demands. These terminals, upon sensing the temperature of the supply air rising above the room temperature, will reduce their supply air volume to the minimum setting or shut off until the rooms requiring heating are satisfied. The system will revert back to the cooling mode after heating is satisfied, and all terminals will revert back to their normal cooling operation.

A warm-up cycle can be built into the system control so that the building can be warmed up prior to its being occupied, after which the internal loads in the building will keep the system in the cooling mode most of the time. Obviously, this system is limited to applications where cooling is the predominant mode of operation. Consequently, this type of system is not suitable for use in cold climates, where a building may have a prolonged heating period, unless a separate heating system is incorporated to neutralize the heat loss along the building perimeter.

Variable volume terminals can be replaced by special diffusers which, in addition to supplying air to the conditioned space, are designed to perform variable volume control function. Most of these variable volume diffusers also incorporate an integral and self-contained thermostat. The major advantage of using this type of diffuser is that each diffuser is a thermostatically controlled thermal zone. A building with many small offices can have one diffuser per room, and each room will have independent temperature control. A cautionary note in the use of these diffusers is that static pressure in the supply duct system must be controlled carefully. For systems serving large buildings with many diffusers, additional pressure control dampers may have to be installed at branch ducts to limit the static pressure in the branch ducts. Otherwise, diffuser noise can become a problem. Many of these diffusers also incorporate a warm-up feature. Upon sensing the temperature of the supply air rising above the room temperature, the diffuser will open fully to warm up the conditioned space quickly.

Variable volume, variable temperature systems are generally used in conjunction with rooftop units. Multiple units are often used, with each unit serving an area of the building with multiple spaces that all have similar, but not the same, load characteristics. To have a large unit serving an entire building with spaces that have different, particularly opposite, exposures may result in control problems.

24 Terminal Reheat Systems

Variable volume terminal reheat systems are the most popular and widely used variable volume systems. This type of system was derived from the constant volume terminal reheat system in an attempt to reduce energy wastage. The schematic diagram of the system, shown in Figure 24.1, is very similar to its constant volume counterpart, shown in Figure 8.1.

The major difference between the constant volume and variable volume versions of the terminal reheat system is that for a constant volume system, the room thermostat controls the reheat coil only, and supply air volume is maintained constant at the terminal at all times. No volume control is required for the supply and return fans in the constant volume system. In the variable volume system, additional controls are added to the reheat terminals. Room thermostats control both the reheat coil and the supply air volume at the reheat terminals. Controls are also added at the supply and return fans to conserve energy and to minimize pressure fluctuations in the duct system as the terminals vary the required volume of supply air at the same time. In the variable volume terminal reheat system, the supply fan and main supply air duct can be sized to deliver a quantity of air corresponding to the block load of the area served by the air handling system, which generally is considerably less than the sum of instantaneous peak air quantities required for a constant volume terminal reheat system.

In a variable volume terminal reheat system, as the space cooling load decreases, the reheat terminal decreases the amount of air supplied to the conditioned space first to maintain the space temperature without activating the reheat coil. When the quantity of supply air is

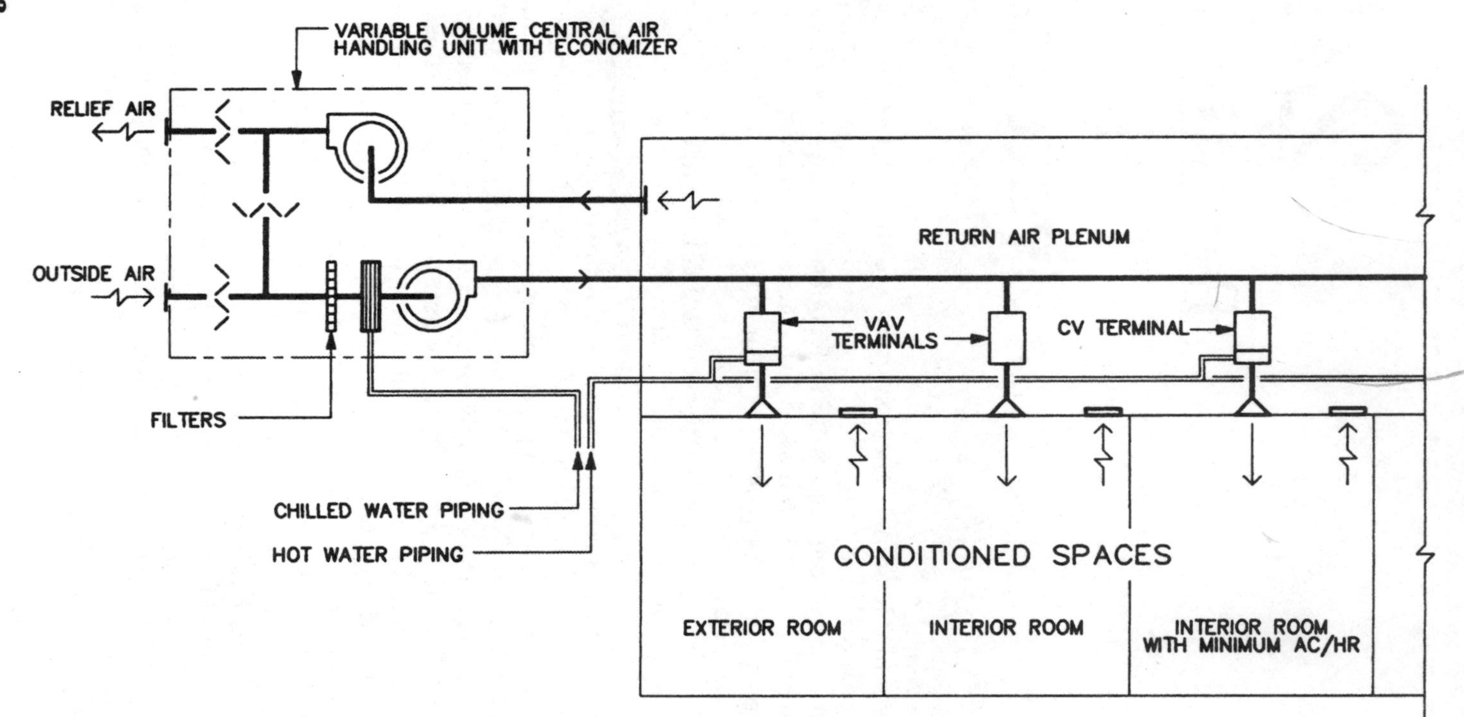

Figure 24.1. Variable volume terminal reheat system.

reduced to a preset minimum, further load reduction will activate the reheat coil. From this point on, performance of the variable volume terminal reheat system is basically the same as that of a constant volume terminal reheat system. Determination of the preset minimum to start the reheat process should take into consideration the adequate air motion needed in the conditioned space, the ventilation requirement, and the maximum supply air temperature under the heating mode.

Reheat energy is the amount of thermal energy needed to heat the supply air from the cold supply air temperature to the room temperature. Reheat energy waste for the variable volume system is substantially reduced compared with the constant volume system simply because the amount of supply air to be reheated is reduced. The amount of reheat energy savings depends on the minimum settings for the various terminals. For each reheat terminal, the hot-water heating coil or electric duct heater is sized to heat the minimum amount of supply air (instead of the total amount of supply air for a constant volume reheat terminal) from the cold supply air temperature to a higher temperature to satisfy the heating load.

Note that since the supply air volume is reduced for heating, the supply air temperature required to satisfy the peak heating load can be substantially higher than that for a constant volume terminal reheat system. If the calculated supply air temperature corresponding to the minimum supply air volume cannot be achieved or is too high for comfort, the minimum setting of the terminal may have to be increased in order to lower the temperature of the supply air. Increasing supply air volume for heating, however, is not a desirable option from an energy conservation point of view because more air will have to be reheated before effective heating can take place.

To conserve energy, the terminal also can be designed such that after supply air volume reaches its minimum, as the load continues to drop, supply air volume can increase again as the heating load increases. This type of terminal will reduce reheat waste during partial load conditions. Figure 24.2 shows various control schemes for a variable volume terminal reheat system.

Variable volume reheat terminals can be arranged so that there is a zero flow point as the supply air volume varies through its flow range. The zero flow point also can be expanded to have an adjustable "dead band" within which no air is supplied to the conditioned space, as shown in case *E* of Figure 24.2. This type of terminal, although energy efficient, should be used with caution, since in an occupied exterior space, the internal heat gains may offset the transmission losses under

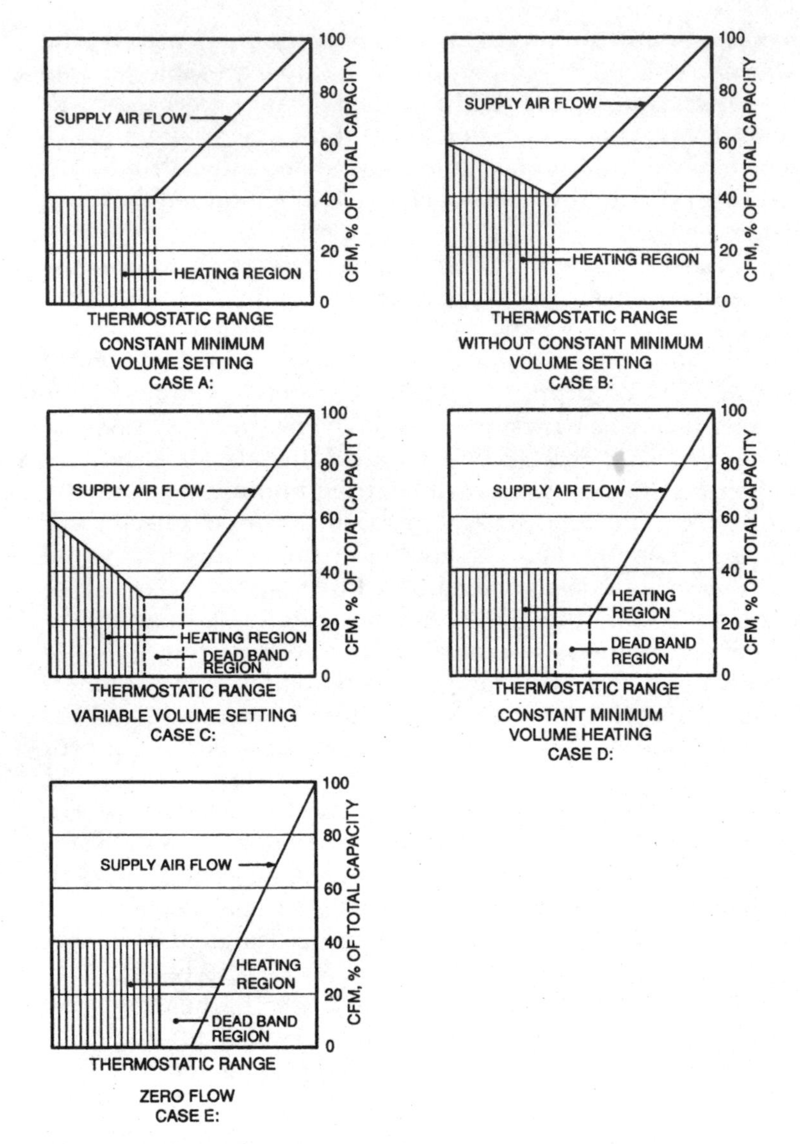

Figure 24.2. Control schemes for variable volume reheat terminals.

partial load heating conditions. Setting a very low minimum amount of supply air into a conditioned space or having a zero flow point in the terminal design, although saving more energy, may cause insufficient air motion in the room and deprive the occupants of required ventilation air for long periods of time.

Only the exterior spaces will require heating. (Heating also may be required for interior spaces in the basement or over an unconditioned parking garage.) Interior spaces require cooling year round. A no-load condition for an interior space always coincides with the unoccupied condition, with the exception of spaces over unconditioned areas or slabs on grade. With the variable volume feature, it becomes apparent that reheat may not be required for interior space temperature control as long as the variable volume terminal can reduce supply air volume to zero under no-load conditions. Therefore, as long as the minimum ventilation rate does not have to be maintained at all times, variable volume terminals without reheat coils may be used for interior spaces.

An economizer cycle is commonly used in conjunction with variable volume terminal reheat systems to conserve cooling energy. Note that with the variable volume feature, reheat seldom occurs in the cooling season unless the minimum supply air volume at the reheat terminal is set very high for a specific reason. During the heating season, the economizer cycle will satisfy the cooling needs by mixing proper amounts of outside air with return air. Mechanical cooling can be shut off. Under this scenario, the word *reheat* becomes a misnomer, since the supply air is never mechanically cooled; thus in strict sense, variable volume terminal reheat systems do not waste "reheat" energy.

From an applications point of view, the variable volume terminal reheat system is a much more flexible air handling system compared with the cooling-only variable volume systems discussed in Chapters 19 through 23. Not only may terminals with or without reheat be used for interior spaces, depending on the space need, either constant volume or variable volume reheat terminals can be used for any conditioned space. For example, a room with a fixed exhaust requirement may need a constant volume reheat terminal in order to maintain positive pressure between the room and its adjacent spaces at all times. Obviously, one must understand that more energy will be used under these conditions.

Dehumidification control for a variable volume terminal reheat system is not as good as that of its constant volume counterpart because the terminal takes care of most cooling load variations in the variable volume mode. Reheating does not take place immediately as the cooling load is reduced from peak condition. For comparison, see the discussion in Chapter 19 of the dehumidification issue for various types of air handling systems.

Some spaces served by a variable volume terminal reheat system may require dehumidification. Under these circumstances, constant volume terminals can be used for dehumidification purposes with the understanding that more energy will be required. Another scheme for achieving dehumidification and also conserving energy is to modify the variable volume terminal control. A humidistat can be installed in addition to the thermostat for spaces requiring humidity control. When the humidity in the conditioned space exceeds a preset limit, a control relay can override the volume control at the terminal to supply the full amount of supply air to the space. The thermostat, in order to satisfy the temperature requirement of the conditioned space, will activate the heating coil to heat the cold supply air to maintain the desired room temperature. In other words, the terminal i s changed from a variable volume to a constant volume reheat operation when dehumidification is required. Note that this control action takes place only when dehumidification is needed. The terminal operates as a normal variable volume terminal under all other partial-load conditions. A schematic diagram and the psychrometric process of this control scheme are shown in Figures 24.3 and 24.4, respectively. It is important to note that if a particular terminal is expected to perform reheat for humidity control, the quantity of supply air for the space must be calculated on the basis of the reheat requirements.

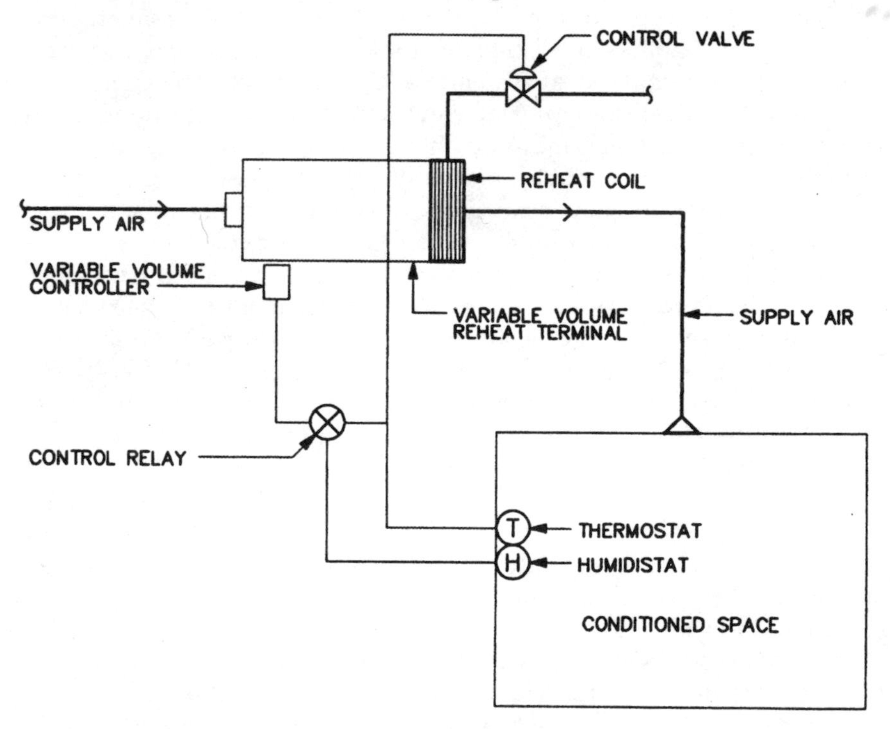

Figure 24.3. A control scheme variation for a variable volume terminal reheat system.

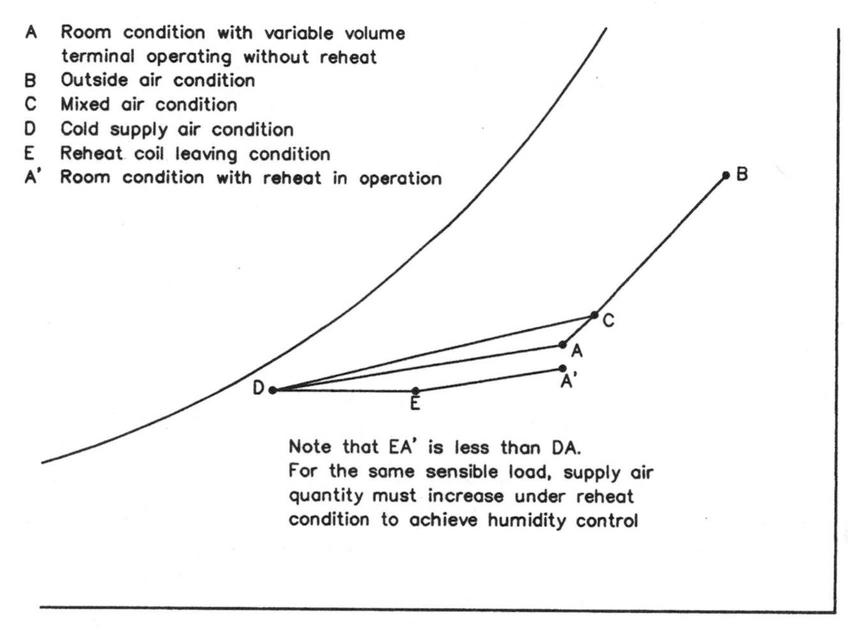

Figure 24.4. Psychrometric process of using a reheat coil for humidity control.

Variable volume terminal reheat systems offer good temperature control and energy savings. These systems are simple to design, easy to install, and require a relatively small fan room for the air handling equipment, and less ceiling space for distribution. These systems are also very versatile and conserve both thermal and transportation energy. Consequently, the variable volume terminal reheat system is often chosen as the model system for energy conservation. Many energy codes and standards choose this system to establish "energy budgets" to be used as maximum allowable energy expenditures for building design.

The disadvantage of the variable volume terminal reheat system is that it is an air-water system. In addition to the sheetmetal ductwork distribution, the system also requires a piping distribution. In the case of electric heat, it requires an electrical distribution, which involves a different trade. With hot-water reheat, there is also a potential water leakage problem with the hot water system in a ceiling space. Compared with the variable volume, variable temperature system, the initial cost of the terminal reheat system can be considerably higher, a price often justifiable, however, by the added flexibility and better control.

25 Double-Duct Systems

The variable volume double-duct system was derived from its constant volume counterpart in an attempt to reduce energy wastage using the same approaches as employed in evolving the variable volume terminal reheat system from the constant volume terminal reheat system. The schematic diagram of the variable volume double-duct system, an shown in Figure 25.1, is very similar to that of the constant volume double-duct system, as shown in Figure 9.1.

The main difference between the constant and variable volume double-duct systems is that volume controls are added to the double-duct terminals for variable volume control. In order to minimize pressure fluctuations in the duct distribution system and to conserve transportation energy, variable volume controls are added to the supply and return fans to deliver only the required air quantity to satisfy the needs.

In a variable volume double-duct scheme, as the space cooling load decreases, the cold air damper in the double-duct terminal will reduce the amount of cold air supplied to the space to satisfy the space cooling load while the hot air damper remains in the closed position. When the volume of supply air drops to a preset minimum, further cooling load reduction will cause the hot air damper to modulate open. As the space load changes from cooling to heating, more hot air is supplied to the space as the cold air damper modulates to its closed position.

Only the exterior spaces where heat loss occurs will require heating (heating also may be required for interior spaces below the roof, in the basement, or over an unconditioned parking garage). These places require terminals with both heating and cooling supplies. The minimum setting of the terminal should take into consideration the ventilation requirements and adequate air motion in the conditioned space.

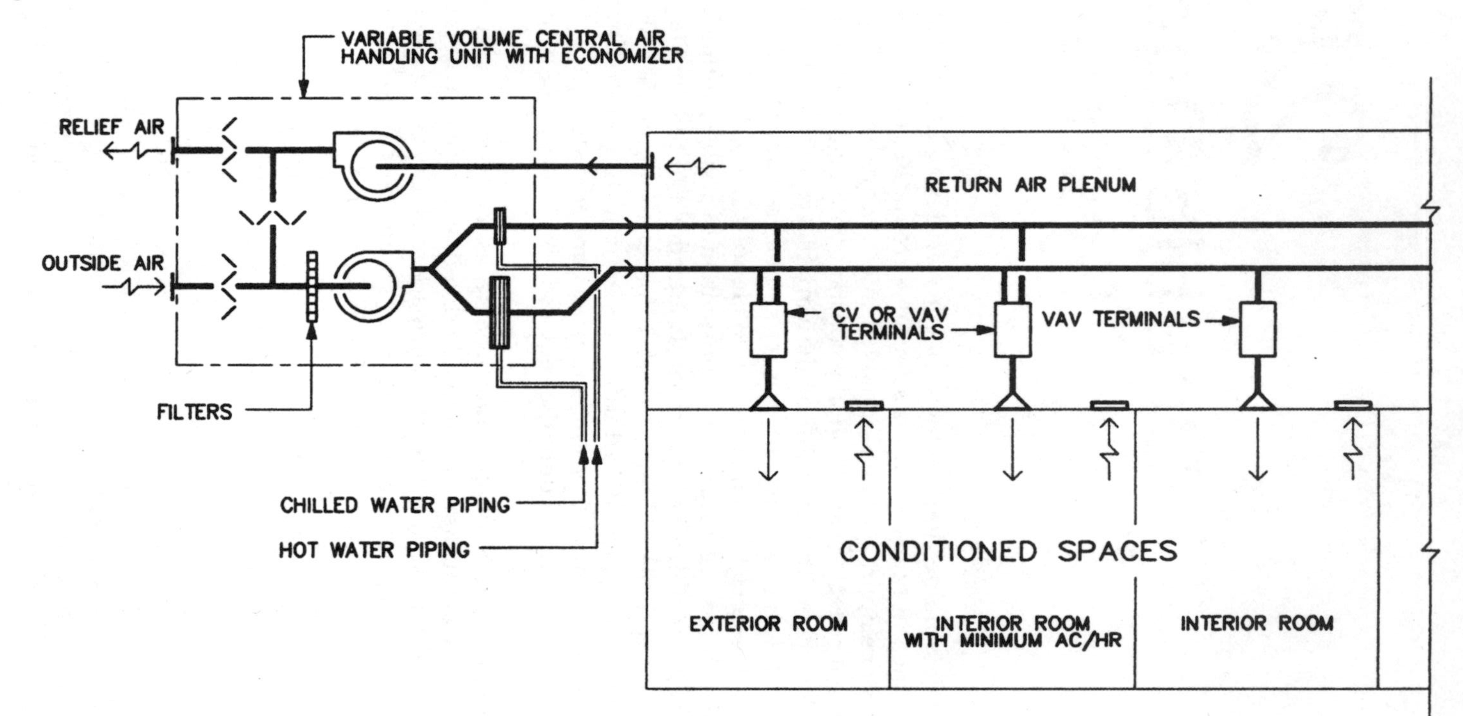

Figure 25.1. Variable volume double-duct system.

There are many different ways to vary the volume of supply air and mix hot and cold air at the terminal to control the space temperature. Figure 25.2 shows several control schemes for varying air volume and mixing air at a variable volume double-duct terminal.

Variable volume double-duct terminals also can be arranged with a

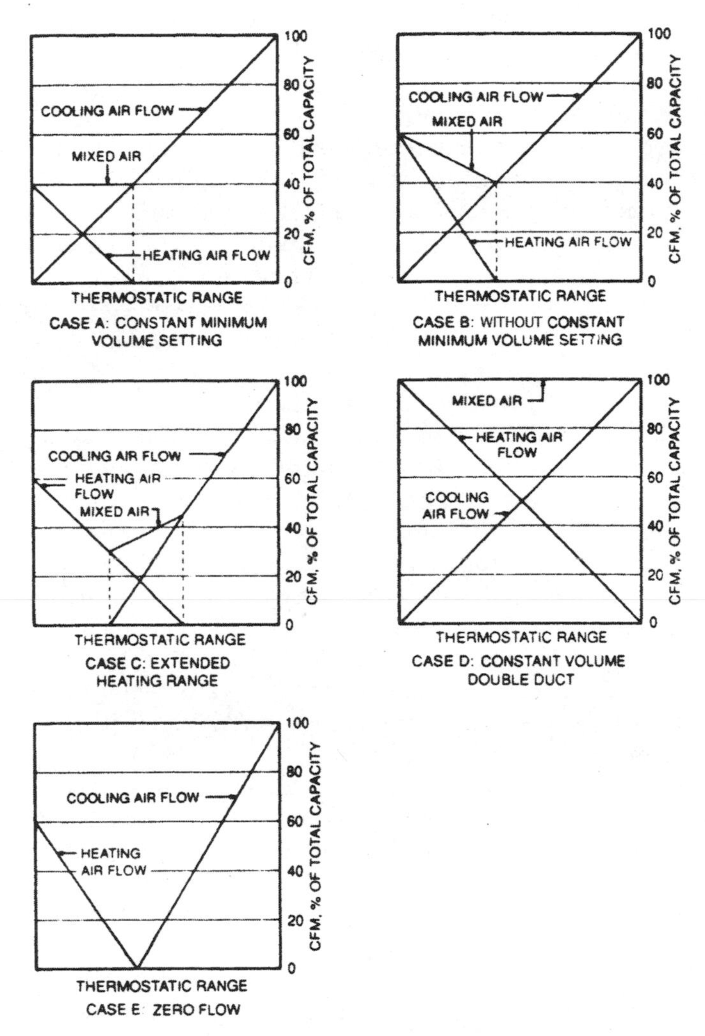

Figure 25.2. Control schemes for variable volume double-duct terminals.

zero flow point as shown in case *E* of Figure 25.2, or a zero flow "dead band." At the zero flow point or within the dead band, neither hot nor cold air is supplied to the conditioned space. The control scheme shown in Figure 25.2 is very similar to that in Figure 24.2 for the variable volume terminal reheat system. One should recognize that for an occupied exterior space, the internal heat gains may offset the transmission losses under partial heating conditions. A terminal with a control dead band should be used with caution, since the no-flow condition can occur when heat loss neutralizes heat gain while the conditioned space is occupied. During these false no-load periods, occupants will be deprived of ventilation air.

Interior spaces require cooling year-round, and the no-load condition for an interior space always coincides with the unoccupied condition, so there is no need for interior spaces to use double-duct terminals with a heating source for temperature control. As long as a cooling-only variable volume terminal can reduce the volume of supply air to zero under no-load conditions, the terminal can satisfy the space needs. And as long as the minimum ventilation rate does not have to be maintained at all times, including unoccupied hours, single-duct cooling-only variable volume terminals may be used for interior spaces.

The variable volume double-duct system is also a very flexible system. Double-duct or single-duct variable volume terminals may be used for interior spaces of different needs. Furthermore, depending on the space requirement, either constant or variable volume double-duct terminals can be used for any conditioned space. For example, a room may require a fixed number of air changes. The quantity of supply air for the required air changes may be equal to or exceed the quantity of supply air required to satisfy the maximum load. In this case, a constant volume double-duct terminal can be used to satisfy the room needs, while other spaces served by the same system can still have variable volume terminals to conserve energy.

The supply fan for a variable volume double-duct system is sized to deliver the air quantity corresponding to the block load of the system, which is generally less than the sum of instantaneous peak air quantities required for a constant volume system. The sizing of the cooling coil and main cold air duct for a variable volume double-duct system is considerably simpler than for a constant volume system. Essentially, the cold air duct can be sized based on the block sensible cooling load of the system. The effect of the loads on the bypass air, as discussed in Chapter 9 for the constant volume double-duct system, should not occur with the variable volume system under peak cooling load conditions. The only exception is that if constant volume terminals or terminals with very high minimum volume settings are used in a variable volume double-duct system, supply fan volume and cooling coil size must be modified to include these influencing factors. One should still keep in mind that in addition to the heating and cooling loads, the design temperatures for the hot and cold supply air also affect the size of the main ducts. See the discussion on this issue for constant volume double-duct systems in Chapter 9.

Size of the supply fan and its motor horsepower are comparable with those of the variable volume terminal reheat system. Again, volume control should be installed at the fan to prevent excessive pressure buildup in the duct system. The variations in volume control schemes for fans mentioned in Chapter 19 also apply to variable volume double-duct systems.

Factory-manufactured terminals are always utilized in variable volume double-duct systems. Field-fabricated control dampers for mixing and volume control are rarely used. Variable volume double-duct terminals, similar to their constant volume counterparts, also have several control schemes. Two single-duct variable volume terminals with mechanical volume control devices can be grouped together to form a variable volume double-duct terminal. The room thermostat controls the supply air volumes from the cold and hot terminals independently to maintain proper room temperature. A more common scheme is the use of a manufactured variable volume double-duct terminal with two reset controllers to control cold and hot damper positions in response to differential pressure sensors located in the cold and hot air inlets at the terminal. The controllers will respond to the room thermostat signal and position the cold and hot air dampers to maintain proper room temperature. As pressures in the cold and hot ducts fluctuate, the differential pressure sensors through the controller will reposition the dampers to compensate for the pressure changes.

This control scheme also can be used to control a terminal that requires constant volume flow in a variable volume system. The total volume of supply air under this type of control, however, may deviate from the constant-airflow requirement during the mixing stage. A more positive way to maintain constant airflow is to have one reset controller controlling the cold air damper to satisfy the room thermostat and another controller maintaining the total supply air volume by controlling the hot air damper in response to a differential pressure signal, with the sensor located in the supply outlet of the terminal. See the discussion in Chapter 9 of control for constant volume double-duct systems.

The psychrometric process of a variable volume double-duct system under cooling mode is the same as that for a variable volume cooling-only system, as shown in Figure 19.3, until the terminal control reaches its minimum volume setting. Under minimum volume conditions, the psychrometric process is the same as that for a constant volume double-duct system, as shown in Figure 9.3. In general, a variable volume double-duct system offers better dehumidification compared with its constant volume counterpart, although the level of dehumidification is not as good as in terminal reheat systems, as shown in Figure 8.2.

Compared with variable volume terminal reheat systems, the double-duct system offers the advantage of being an all-air system. There is no piping or additional electrical distribution system, only sheetmetal is used in the double-duct distribution system. There is no concern about potential water leakage through the ceiling. Tenant modification and retrofitting work is easier, since there is only one sheetmetal trade involved. However, a double-duct system will require a larger fan room, more sheetmetal, and more ceiling space for duct distribution. Consequently, construction costs for a variable volume double-duct system are generally higher than those for variable volume terminal reheat systems.

Energy waste for variable volume double-duct systems is reduced compared with constant volume double-duct systems simply because the amount of mixing between cold and hot air is reduced. The amount of energy savings depends on the minimum setting of the variable volume double-duct terminal. Again, one should be reminded that setting a very low minimum amount of supply air into a conditioned space, although saving more energy, may have adverse effects on the occupants. With the use of single-duct cooling-only terminals for the interior spaces, energy waste for terminals serving interior spaces can be eliminated completely. There is very little difference in annual energy consumption between the variable volume double-duct system and the variable volume terminal reheat system. Since introduction of the variable volume double-duct concept, other variations in double-duct design have become available which, compared with their respective constant volume counterparts, will offer more energy conservation potentials.

The variable volume double-duct system should not be confused with the single-fan dual-conduit system without perimeter cooling, as discussed in Chapter 21. The components used in the air handling unit for these two systems are identical. The design philosophies of the two systems, however, are completely different. The applications, energy use, and system limitations are also different for these two systems.

26 Double-Duct Systems with Subzone Heating

The variable volume double-duct system with subzone heating is a derivation of the variable volume double-duct system and is very similar to the constant volume double-duct system with subzone heating. The air handling system is identical to the constant volume double-duct system with subzone heating except that variable volume terminals are used and the supply fan has volume control. Cold and bypass air main ducts and the mixed air terminals for this system are essentially the same as those in the conventional double-duct system. For exterior zones or where heating is needed for other reasons, small heating coils or electric duct heaters are installed downstream of the variable volume double-duct terminals to provide heating. Heating coils need not be installed for zones serving interior spaces where cooling is required year round. Furthermore, since the system is a variable volume design, single-duct cooling-only terminals can be used for interior spaces as long as the minimum ventilation rate does not have to be maintained at all times. A schematic diagram of such a system is shown in Figure 26.1. Hot-water heating coils are shown in the figure. However, electric duct heaters are often used for these systems, similar to those used in the constant volume version shown in Figure 10.1. Since the bypass air main duct does not carry tempered air, the duct need not be insulated.

The variable volume double-duct system with subzone heating uses

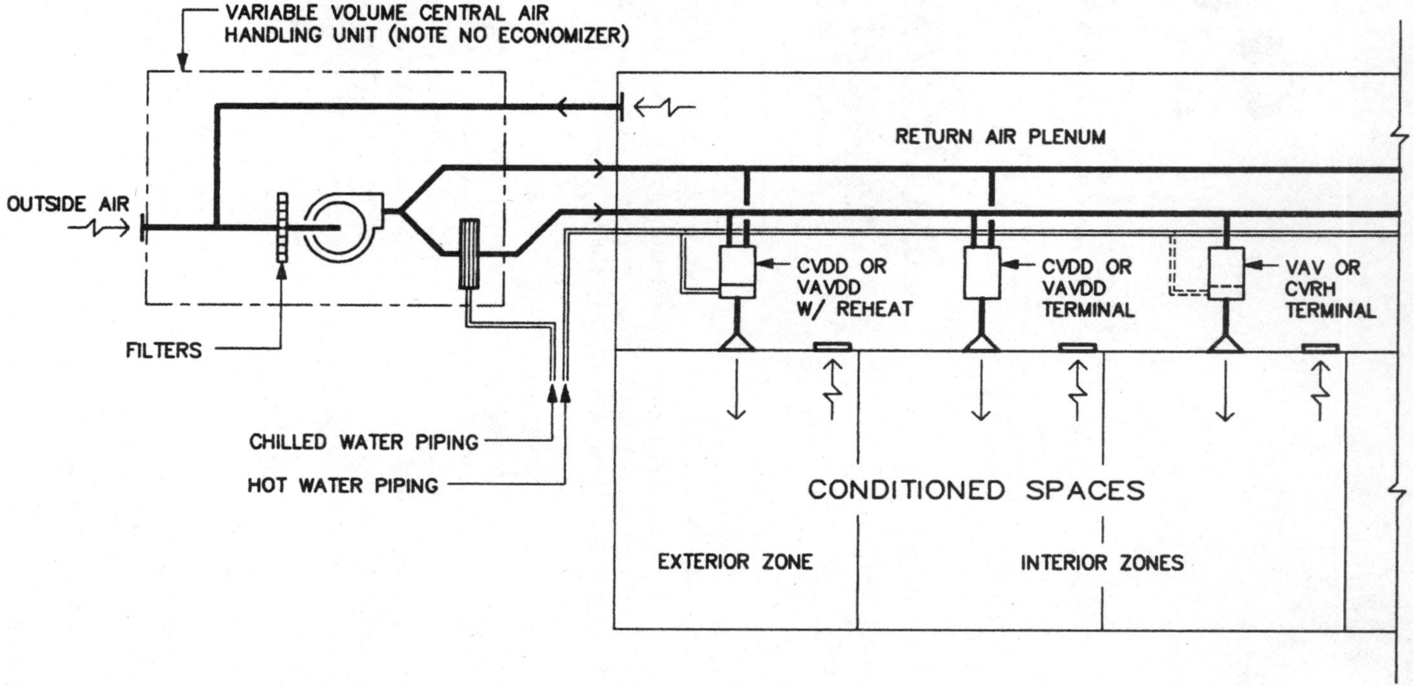

VARIABLE VOLUME CENTRAL AIR HANDLING UNIT (NOTE NO ECONOMIZER)
OUTSIDE AIR
FILTERS
CHILLED WATER PIPING
HOT WATER PIPING
RETURN AIR PLENUM
CVDD OR VAVDD W/ REHEAT
CVDD OR VAVDD TERMINAL
VAV OR CVRH TERMINAL
CONDITIONED SPACES
EXTERIOR ZONE
INTERIOR ZONES

Figure 26.1. Variable volume double-duct system with subzone heating.

the least amount of energy of the variable volume systems. The system is also the most versatile in satisfying a variety of functional requirements. However, the initial cost of the system is among the highest of the variable volume systems. Most of the pros and cons discussed in association with the constant volume double-duct system with subzone heating also apply to the variable volume version.

Use of an economizer cycle with this system should be evaluated carefully. As discussed in Chapter 10, the economizer can only try to satisfy the cold air temperature by mixing more cold outside air with return air. Without a heating coil in the bypass air plenum, the bypass air duct and cold air duct will have the same air temperature, and any terminal that is in the mixing mode will lose control.

A heating coil can be placed in the bypass air plenum to heat the air leaving the economizer to the room temperature. In this case, however, more construction costs for piping and control will be incurred for the added heating coil in addition to the cost of the economizer.

The psychrometric process for this system in the cooling mode is the same as that for the variable volume cooling-only system shown in Figure 19.3 until the terminal reaches its minimum volume setting. Beyond that, the terminal behaves like a constant volume double-duct system, as also shown in Figure 9.3. Under heating conditions, the heating coil at the terminal serving an exterior space will be activated after the cold air damper is closed completely.

Basically, only single-duct variable volume terminals are needed for serving interior spaces. Double-duct variable volume terminals can be used if an interior space requires minimum ventilation at all times. Subzone heating also can be added for dehumidification control if necessary. In this case, when the humidity in the conditioned space exceeds a preset limit, a room humidistat, through a control relay, can reverse the mixing damper position to allow cold air instead of bypass air into the space. The thermostat, in order to satisfy the temperature requirement of the conditioned space, will activate the heating coil to heat the cold supply air and maintain the desired room temperature. The amount of supply air can be varied if necessary. For simplification and quick response to the dehumidification need, however, the cold air supply damper should be set to the fully open position when the room humidistat calls for dehumidification.

It is important to note that if reheat is a part of the original design for a particular space under full load conditions, the supply air quantity for the space requiring dehumidification control must be calculated on the basis of the reheat requirement. The schematic control diagram and the psychrometric plot for this arrangement are similar to those shown in Figures 10.2 and 10.3, respectively, for the constant volume system.

Single-duct variable volume reheat terminals also can be used for this system, as shown in dashed lines in Figure 26.1. Reheat terminals can be used to provide minimum ventilation if required or for dehumidification. For an interior space, it is interesting to note that it is more energy efficient to use a variable volume double-duct terminal to maintain minimum ventilation requirements, and it is easier and simpler to use a single-duct variable volume reheat terminal for dehumidification applications.

The higher initial and maintenance costs of the variable volume double-duct system with subzone heating, coupled with the least energy use and greater flexibility, can be viewed as a classic case of the need to evaluate the balance between higher initial system cost and more energy savings, and between the convenience of the system's flexibility and the complexity of system maintenance. In addition to life-cycle costing, many intangible factors need to be considered, not the least of which is the analyst's mind set. If the analyst is an energy-conservation fanatic, he or she can find ways to skew the analysis in favor of using this system. On the other hand, if the analyst is initial cost conscious or more concerned about maintenance difficulties, the analysis can easily be skewed in favor of using other types of air handling systems.

27 Double-Duct Systems with Draw-Through Cooling

A variable volume double-duct system with draw-through cooling is a simple modification of its constant volume counterpart. As discussed in Chapter 11, the primary reason for using a draw-through cooling system is for better humidity control. Since the constant volume double-duct system with draw-through cooling wastes a large amount of energy, the variable volume version is a system modification designed to minimize energy waste also.

The differences between the constant and variable volume versions of the draw-through cooling system are the same as those between conventional double-duct systems. Controls are added to the double-duct terminals for variable volume control. Supply and return fans also have variable volume controls to minimize pressure fluctuations in the main duct distribution system and to conserve energy. Single-duct variable volume terminals can be used for interior spaces where heating is not required. Constant volume double-duct terminals also can be used for interior spaces where a minimum ventilation rate must be maintained at all times and where dehumidification control is required throughout the load range. A schematic diagram of this arrangement with an economizer cycle is shown in Figure 27.1.

Other pros and cons of this system are essentially the same as those for the constant volume double-duct system with draw-through cooling (see Chap. 11). The variable volume version makes the double-

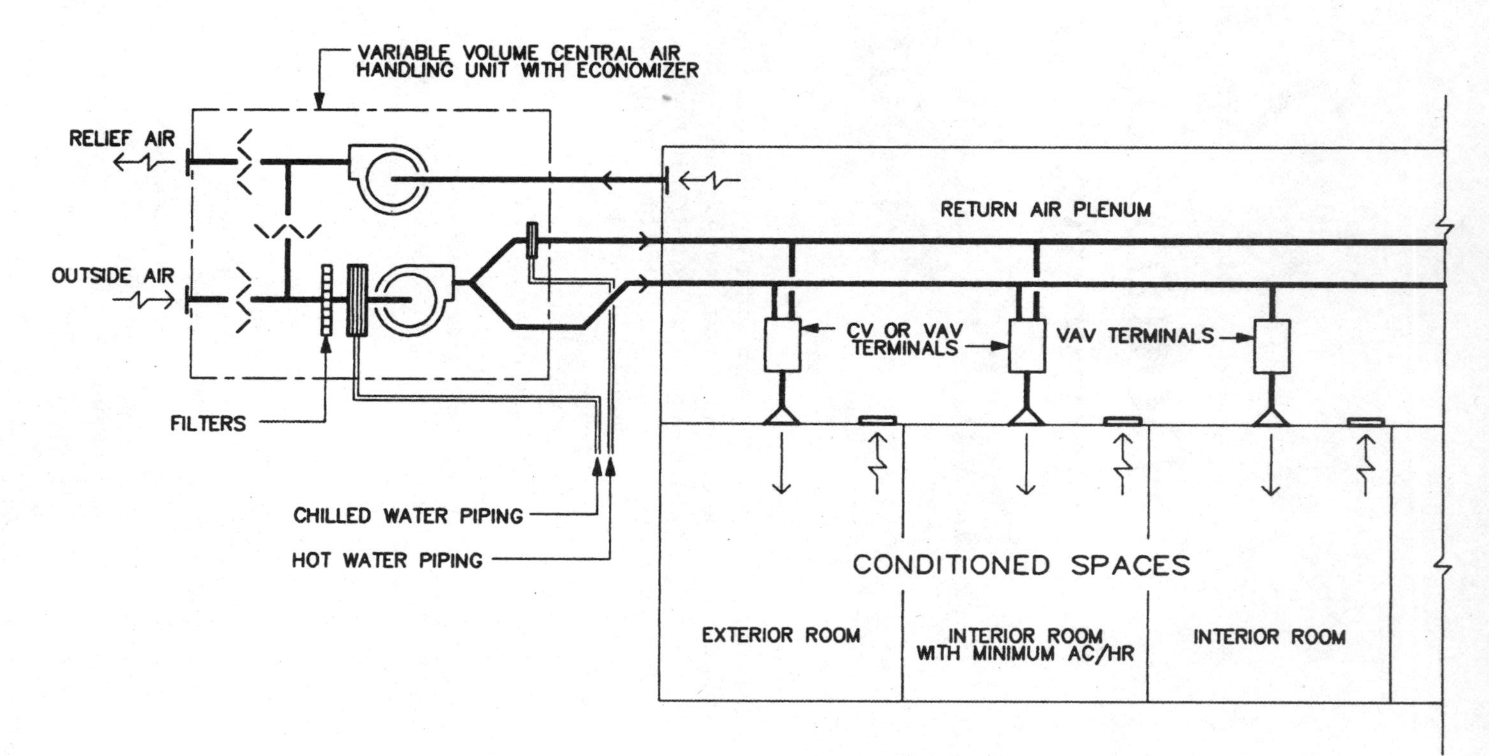

Figure 27.1. Variable volume double-duct system with draw-through cooling.

duct system with draw-through cooling more practical because energy waste is drastically reduced, much the same as the relationship between constant and variable volume versions of the terminal reheat system (see Chaps. 8 and 24). Construction costs for this system are still high when compared with the variable volume terminal reheat system, which also offers good humidity control.

28
Two-Fan Double-Duct Systems

When an economizer cycle is used in conjunction with a double-duct system, more heating energy is wasted because the air leaving the economizer must be heated to the room temperature before being useful. Energy waste can be eliminated if the air handling unit can be designed so that the economizer cycle applies only to the cold supply air and the air that goes through the heating coil is 100 percent return air or return air with a fixed minimum amount of outside air.

To accomplish this goal, the air handling unit must have two supply fans, one for hot air and one for cold air, together with a return air fan associated with the cold supply air fan. When applying this concept to a constant volume double-duct system (see Chap. 12), since the amounts of hot and cold supply air vary at all times, instead of using one constant volume supply fan, both hot and cold supply air fans, together with the return air fan for the economizer cycle, will require variable volume control. The additional equipment and control, together with the required calibration and added maintenance, make the constant volume version difficult to justify.

For a variable volume system, however, since the supply and return fans must have variable volume control to perform the volume control function, all that is needed for a two-fan application is to add a hot air supply fan with variable volume control. Compared with the constant volume two-fan double-duct system for a project where the variable volume system is applicable, the variable volume version of the two-fan double-duct system is a much more reasonable solution from the point of view of energy conservation. A schematic diagram of this system is shown in Figure 28.1. Note that there is no reason to incorporate an

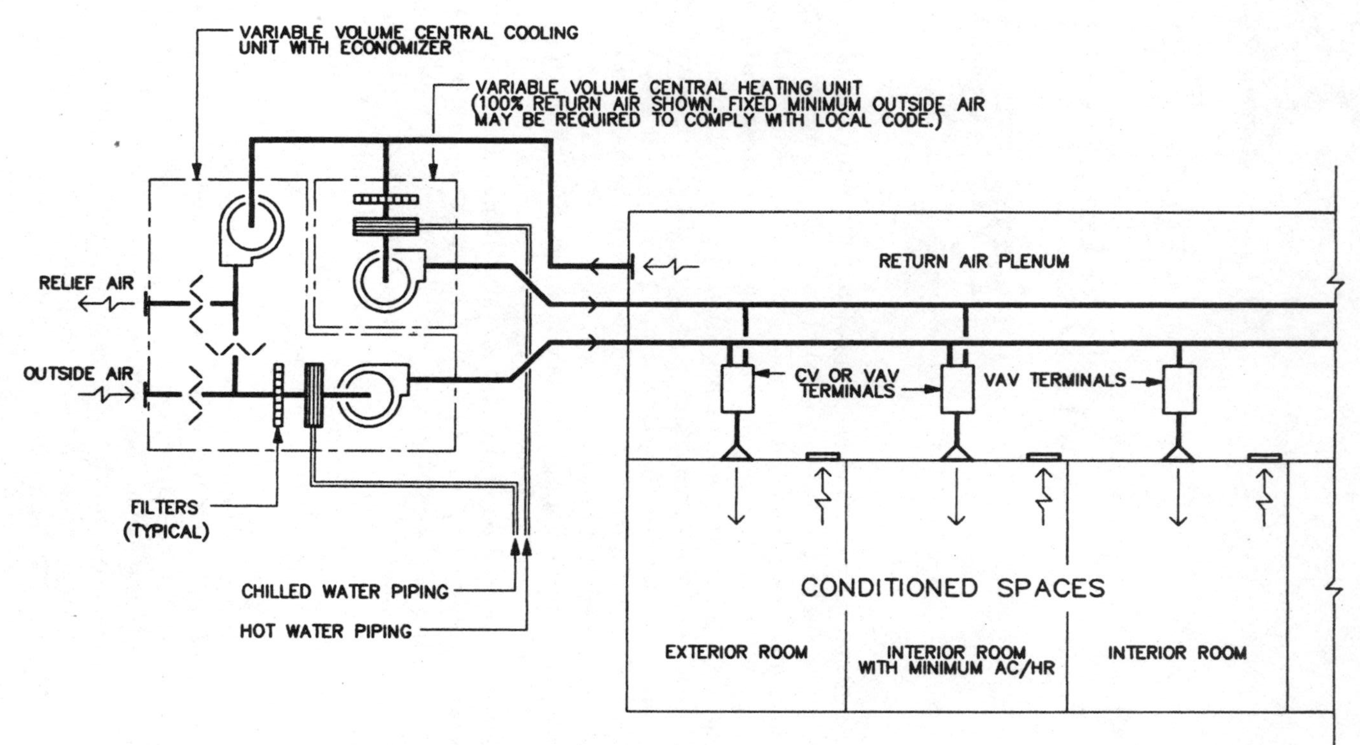

Figure 28.1. Two-fan variable volume double-duct system.

economizer cycle into the heating side. An economizer can only utilize cooler outside air to minimize cooling energy. The purpose of this two-fan system is to separate the heating from the cooling side to avoid the waste of heating energy associated with economizer cycle operation. It is evident that an economizer serves no purpose on the heating side. The duct distribution and mixing terminals of this system are identical to those of the conventional variable volume double-duct systems.

The sole reason for using two supply fans for a double-duct system is to conserve energy. From an energy use point of view, it is desirable to use 100 percent return air through the heating side. On the surface, there seems to be a concern that exterior spaces that demand heating may not get enough ventilation air during the heating season. In reality, since the heating load calculation rarely takes into account heat gains from lights and people, and since the maximum hot air temperature is usually set higher than the temperature required for the worst heating case, at practically no time will any room demand 100 percent supply air from the hot air duct. With proper selection of terminal type, case *A*, *B*, *C*, or *D* in Figure 24.2, ventilation air can always be supplied through the cold air duct under minimum flow conditions. In fact, when exterior spaces require heating, the economizer cycle will be in operation, and more outside air will be introduced through the air-conditioning system. Often, however, this rationale cannot satisfy the increasing concern for IAQ that has surfaced in recent years. Thus some two-fan double-duct systems use fixed minimum amounts of outside air on the heating side to satisfy minimum outside air requirements.

The variable volume two-fan double-duct system also minimizes transportation energy. With a two-fan system, each fan operates independently as a variable volume supply fan. With proper volume control strategy and devices, each fan will use minimum power to deliver the amount of air required to satisfy the system needs at any given time.

The cost of a variable volume two-fan double-duct system is higher than that of a single-fan double-duct system. The only reason for using this system is to save energy. A constant volume two-fan double-duct system saves thermal energy only, whereas the variable volume version saves transportation energy as well as thermal energy. From an energy-conservation viewpoint, it is much more reasonable to use the two-fan system in a variable volume application as compared with a constant volume application. The amount of thermal and transportation energy savings from using a two-fan variable volume double-duct system can be significant, and the energy cost savings may show a reasonable payback period to recover the higher initial costs of the system.

29 Fan-Powered Terminal Systems

One of the drawbacks of variable volume systems is the lack of air motion in the conditioned space during low-load periods. The primary purpose of the fan-powered terminal system is to overcome this deficiency. The central air handling unit for a variable volume fan-powered terminal system is identical to that for a pure variable volume system or a variable volume terminal reheat system. The only difference is that with this system, a small cabinet fan is added at the variable volume terminal. The variable volume portion of the terminal operates the same as that for a pure variable volume system. Under low-load conditions, the terminal fan draws return air from the ceiling plenum, together with variable volume air provided through the central air handling system, and supplies more air into the conditioned space. The fan-powered terminal system also can be viewed as a fan coil unit system integrated into a central variable volume system.

Depending on the type of fan-powered terminal, the terminal fan operates intermittently or continuously to provide adequate air movement in the conditioned space. A schematic diagram of a fan-powered terminal system is shown in Figure 29.1. There are two types of fan-powered terminals: parallel flow and series flow. In a *parallel-flow terminal,* the terminal fan is located outside the central variable volume supply, as shown in Figure 29.1. In a *series-flow terminal,* the terminal fan draws a mixture of plenum return air and variable volume supply air from the central system and supplies the mixed air to the conditioned space.

Parallel-flow terminals are used more commonly than series-flow terminals. The parallel-flow terminal fan runs intermittently, which

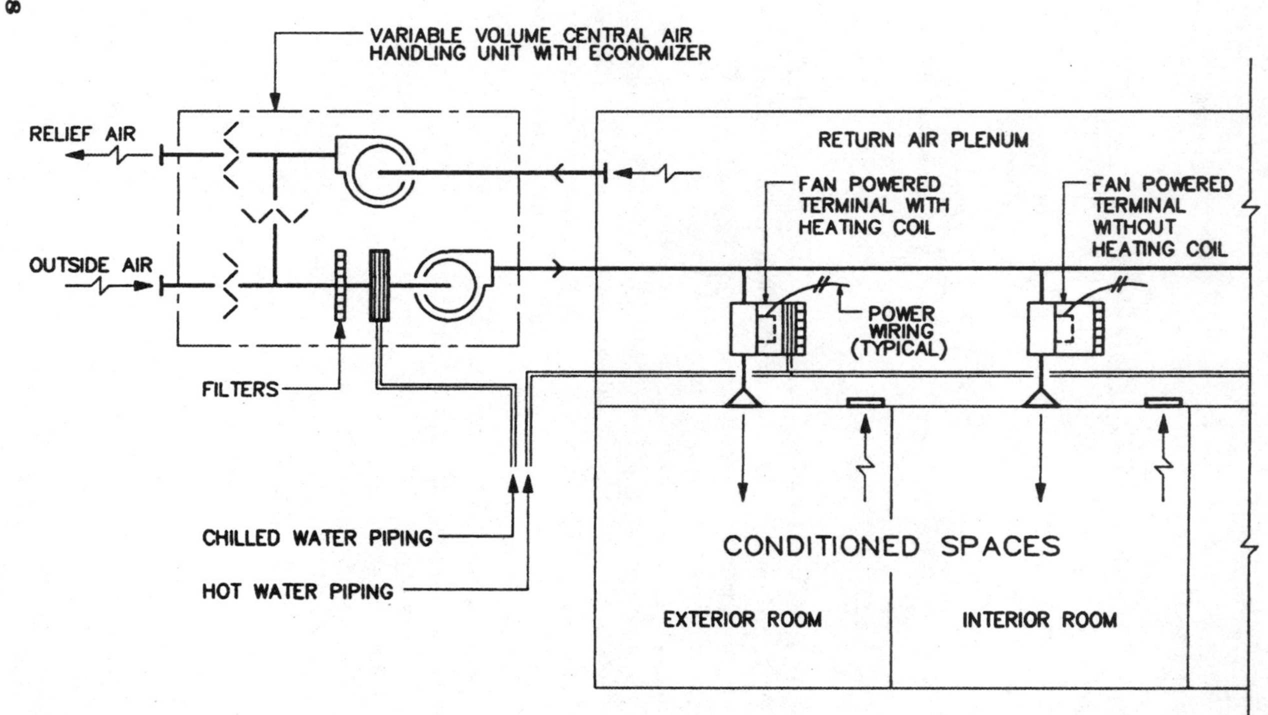

Figure 29.1. Variable volume system with fan-powered terminals.

consumes less transportation energy. With this type of terminal, as the space cooling load decreases, the terminal decreases the amount of air supplied to the conditioned space to maintain the space temperature, much the same as a pure variable volume system. At this point, the terminal fan is not operating. When the volume of supply air drops to a preset limit, further load reduction will activate the terminal fan. From this point on, air motion in the conditioned space will be maintained by the terminal fan, and the room thermostat continues to control the amount of supply air required from the central variable volume system.

As stated in Chapter 4, the temperature of the return air in the ceiling plenum is often higher than the room temperature because various load components will give out heat to the ceiling cavity as well as to the conditioned space. As the terminal fan is activated, the higher-temperature return air will be introduced into the conditioned space by the terminal fan, and the heat to return air will appear as added load to the conditioned space. This added heat tends to slow down the variable volume reduction and provide more ventilation air into the room. In cases where the temperature of the return air is considerably higher than the room temperature, the heat from the return air will actually provide some degree of heating to the conditioned space.*

For terminals serving exterior spaces, a heating coil can be installed at the terminal fan to add more heating to the space, as shown in Figure 29.1. In this case, the terminals act like variable volume reheat terminals except that with this system the heat is added to the return air through the terminal fan instead of to the minimum supply air quantity of a variable volume reheat terminal. With the heating applied to a larger volume of return air, the minimum supply air quantity of the variable volume terminal is no longer influenced by the supply air temperature limitation (see discussion of the variable volume terminal reheat system in Chapter 24), and the amount of reheat is minimized.

Series-flow terminals can be applied to spaces where near-constant air movement is needed at all times. With this type of terminal, air movement in the conditioned space is maintained above a level determined by the capacity of the terminal fan. Variable volume supply air from the central air handling system is added on top of this maintenance level. Heating coils also can be used with this type of terminal

*It should be noted that temperature in the return air plenum may not be as high as anticipated in calculations. In many instances, heat loss through the supply air duct and duct leakage can lower the anticipated return air temperature.

serving exterior spaces. The disadvantage of this type of terminal is that the terminal fan has to operate at all times, thus using more transportation energy.

Fan-powered terminals have a distinct advantage in morning warm-up over other variable volume systems. The central air handling system need not be turned on during the warm-up period. Small cabinet fans with and without heating coils can be used to circulate plenum air to the conditioned space to warm up the building. With lights turned on, the warmer air in the ceiling plenum can be circulated to the conditioned space for the interior spaces where no heating coil is used in the fan-powered terminals.

Since the fan-powered terminals rely on the use of return air in the ceiling cavity, the system may be difficult to apply where ducted return is used. With the added terminal fans, additional electrical wiring will have to be extended to all fan-powered terminals. Additional maintenance time also will be required to service these terminals. Acoustics can become a concern with fan-powered terminals because of the added fan noise. This is particularly true with parallel-flow terminals, in which the terminal fans operate intermittently. In effect, besides the fact that fan-powered terminals serving interior spaces may not require a heating coil and that all cooling is performed at the central air handling unit, there is little difference between this system and a fan coil unit system with a central outside air system.

The plenum return air is "used" air. Filters should be installed at the fan-powered terminals to remove dirt particles and contaminants from the return air before supplying it into the conditioned space. The drawback is that the small fans used in these terminals cannot be sized to accommodate the added static pressure drop across higher-efficiency filters without fan noise becoming a pronounced problem and the low-efficiency throwaway-type filters may not be sufficient to meet the IAQ requirements. Furthermore, because these terminals are always installed in the ceiling space, filters, if installed, are rarely maintained. Indoor air quality thus can become a problematic issue with this type of system.

30 Double-Duct Systems with Extra Cooling Coils

This is a rarely used variable volume double-duct variation. The system is mentioned here primarily to demonstrate that unique system variations can be generated and tried for special applications. With the development of terminal devices and control technology, this system may become a member of the variable volume double-duct family.

Because of the nature of double-duct system operation, the main hot and cold air ducts are each sized for more than 50 percent of the total supply air requirement. This is particularly true for the cold air duct, which is rarely sized for lower than 80 percent of the required total flow. The sum of cross-sectional areas of the cold and hot air ducts is generally equivalent to approximately 130 to 160 percent of total airflow. These larger ducts reflect as additional sheetmetal costs and become a drawback of the double-duct system that should be minimized if at all possible. At the same time, larger ducts also take up more ceiling space and create more congestion in the ceiling space. This system variation is aimed at minimizing these deficiencies.

A conventional double-duct system can be modified by adding a cooling coil in the hot plenum in series with the heating coil, as shown in Figure 30.1. In a variable volume double-duct system, since no heating is needed in the cooling season and the heating coil is always deactivated, the hot duct is practically useless during that time. By adding a cooling coil in the hot plenum and activating it during the cooling

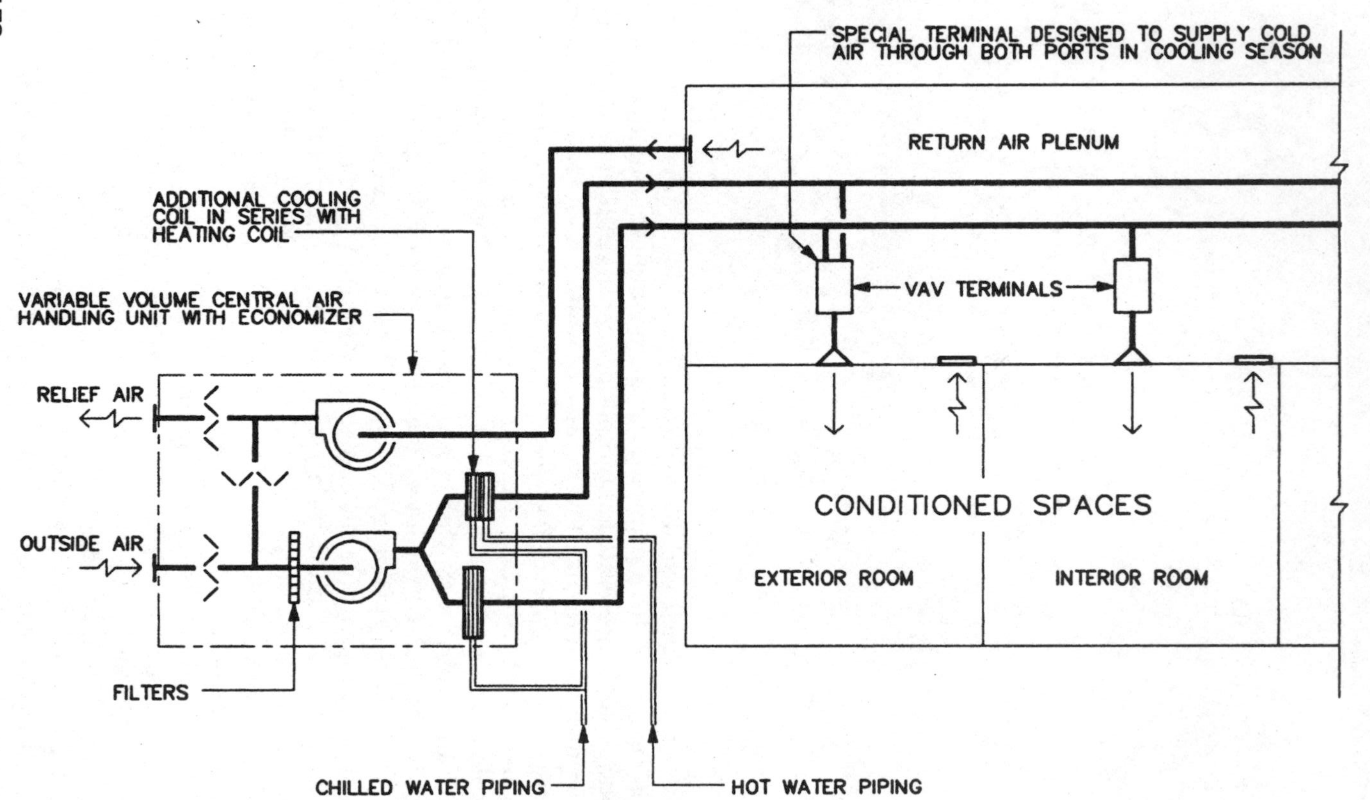

Figure 30.1. Variable volume double-duct system with an additional cooling coil.

season, the hot air duct can be converted into another cold air duct to help deliver cold air. Obviously, with both ducts supplying cold air, the size of the original cold air duct can be reduced. It is conceivable that the sum of the cross-sectional areas of both the hot and cold ducts can be reduced to handle just 100 percent of the total flow, a significant 30 to 60 percent size reduction.

With this concept, however, operation of the conventional double-duct terminals must be modified. During the heating season, the terminals for this system will operate exactly the same as variable volume double-duct terminals. During the cooling season, however, the terminal control will have to be switched so that both hot and cold ports at the terminal can deliver cold air. Instead of operating a cold air actuator only during the cooling season, both actuators will operate in tandem under peak cooling conditions. This was a more difficult task with earlier designs of variable volume double-duct terminals. With newer terminal designs, in which separate controllers activate hot and cold actuators, and with the advent of direct digital control (DDC) technology, terminals can be readily converted to perform this type of control modification.

With this system, interior spaces that have cooling needs year round will use single-duct variable volume terminals. It should be noted that with a conventional double-duct system, any variable volume double-duct terminals can have a minimum setting for the required minimum circulation rate. When the load drops below the minimum setting, varying amounts of warm air will mix with cold air to maintain the minimum volume required and also satisfy the load requirement. Constant volume terminals also can be used if critical pressure relationships between adjacent spaces must be maintained. With this modification, however, since the double-duct system will perform like a variable volume cooling only system during the cooling season, the options mentioned above will no longer be available. Giving up these attractive features normally available for a conventional variable volume double-duct system can be a major drawback of this system. The engineer should weigh these options against the sheetmetal savings or consider other variable volume system options before making a final decision.

The double-duct system with an extra cooling coil should not be confused with the single-fan dual-conduit system discussed in Chapter 21. The components used in the air handling unit are identical for these two systems, but the design philosophies of these two system are completely different.

PART 4

Other System Considerations

31 Hydronic Heat Pump Systems

A *heat pump* is a refrigeration cycle that absorbs heat from a lower-temperature source and elevates the absorbed heat to a useful level and supplies it to the conditioned space for heating. Unitary heat pumps generally are designed so that they can perform cooling duty in the cooling season and produce heat in the heating season. By definition, a heat pump is a primary air-conditioning system. While this book is devoted to the discussion of secondary air-conditioning systems, basically air handling systems, a hydronic heat pump system is a unique decentralized primary-secondary system combination that deserves some discussion.

A hydronic heat pump system, sometimes referred to as *closed-loop water source heat pump*, is a series of heat pumps linked together by a piping loop, with water circulating in the loop forming a reservoir that acts as a heat sink as well as an artificial heat source. Originally, water source heat pumps were developed using groundwater as heat source or sink. Since these heat pumps required mild-temperature groundwater to function efficiently and mild-temperature groundwater does not exist everywhere, the closed-loop water source heat pump system was developed in the early 1960s in which an artificially created reservoir is used as the heat sink as well as the heat source. A typical arrangement of a hydronic heat pump system is shown in Figure 31.1.

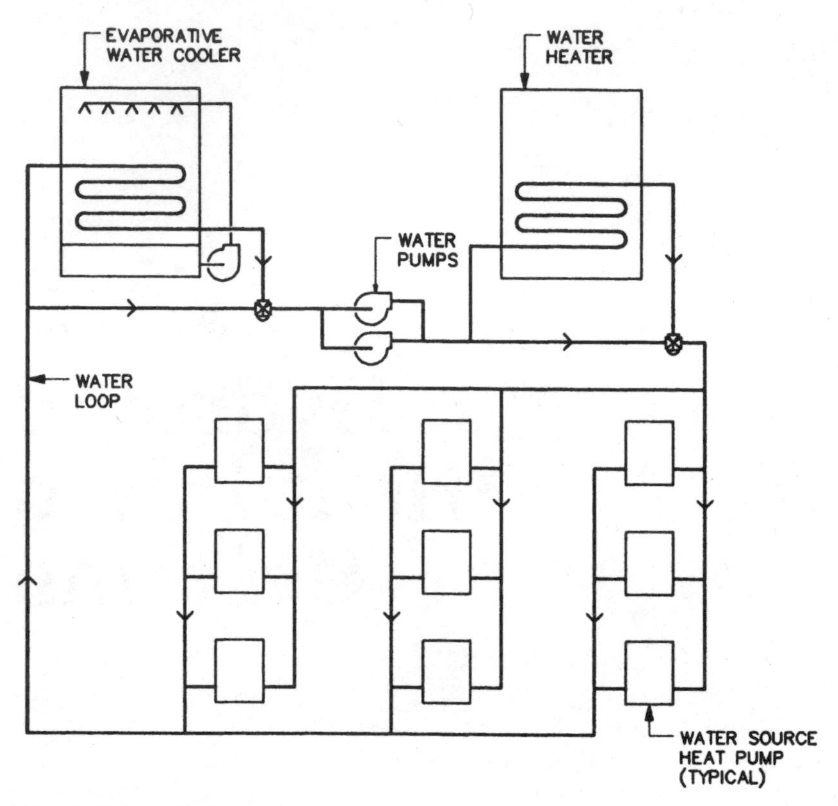

Figure 31.1. Hydronic heat pump system.

Individual heat pumps connected to a common piping loop perform cooling and heating functions in response to the space needs. Heat pumps in the cooling mode will absorb heat from the conditioned space and "reject" the heat into the circulating water loop, whereas heat pumps in the heating mode will absorb heat from the water loop and "pump" the heat into the conditioned space. A typical water source heat pump operation in cooling mode and heating mode is shown in Figure 31.2.

During the cooling season, all heat pumps reject heat into the circulating water loop. As water temperature in the loop rises, the rejected heat can be removed through an evaporative water cooler installed in the water loop. During the heating season in a cold climate, most heat pumps will be in the heating mode, in which they absorb heat from the circulating water in the loop. As the water temperature in the loop falls below a certain limit, a water heater installed in the loop heats the

BLOWER
COIL
CONDITIONED AIR
CAPILLARY TUBE
REVERSING VALVE SHOWN IN COOLING MODE
COMPRESSOR
WATER TO REFRIGERANT HEAT EXCHANGER
CENTRAL WATER LOOP SUPPLY
RETURN

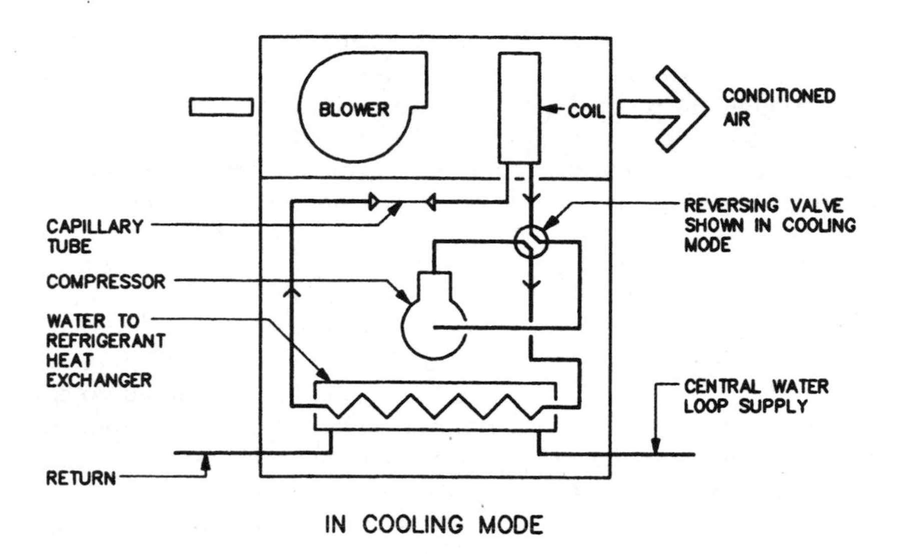

IN COOLING MODE

BLOWER
COIL
CONDITIONED AIR
CAPILLARY TUBE
REVERSING VALVE SHOWN IN HEATING MODE
COMPRESSOR
WATER TO REFRIGERANT HEAT EXCHANGER
CENTRAL WATER LOOP SUPPLY
RETURN

IN HEATING MODE

Figure 31.2. Hydronic heat pump operation.

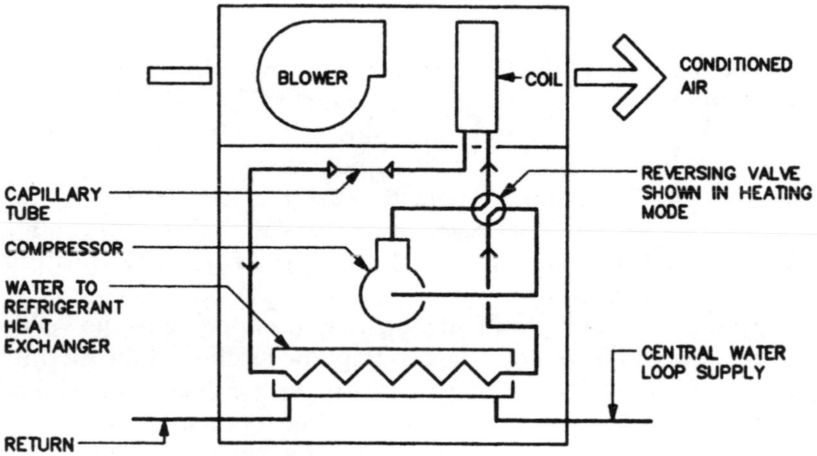

water to maintain the minimum water temperature. In moderate weather, during which some of the units require heating while others require cooling, the rejected heat from the cooling units can be used as the heat source for the heating units.

It is interesting to note that a building generally has some cooling needs year-round. During the heating season, a certain amount of rejected heat is almost always available to help provide heat for heating. In the spring and fall, it is conceivable that for periods of time neither the water heater nor the evaporative water cooler is needed, and the heat pumps provide heating and cooling to each other to sustain system operation. Storage tanks can be installed in the water loop to increase the "reservoir" capacity and prolong the period of this comrade relationship among the heat pumps. Solar heat collectors also can be used in conjunction with this system to provide an additional heat source.

The initial installation cost of a hydronic heat pump system is relatively low. This is particularly true for a building with lease tenants, since only the closed-circuit water loop together with the water heater and evaporative water cooler need to be installed with the shell of the building. A few heat pumps may have to be installed initially for the public core spaces of the building. Most of the heat pumps for tenant areas can be installed later when and as the tenant spaces are leased.

Centralized air-conditioning systems generally operate less efficiently under partial- or low-load conditions. With hydronic heat pumps, individual units in the unoccupied spaces can be turned off without affecting the central system. Because the cooling and heating units share the reservoir, heat rejected to the water loop from cooling units can be used for spaces requiring heating. Energy use with this system is also minimized.

Zoning can be difficult with hydronic heat pump systems, however. Unless each space is served by a heat pump, the location of the thermostat can become a problem (see Fig. 2.2). An electric duct heater can be added to a branch duct that extends from the heat pump to form a subzone. Application of this type of subzone control, however, is limited.

One of the drawbacks of this system is that with a refrigeration compressor in a heat pump unit located in or near a tenant space, noise and vibration can be a problem. With this system, refrigeration equipment together with air handling units are decentralized. Therefore, more maintenance will be required. To alleviate this problem and to minimize field repairs, sometimes it is easier to remove and replace a failed unit with a spare unit. A closed-circuit evaporative water cooler should always be used in lieu of a cooling tower for rejecting heat in the circulating water loop to keep the water loop clean.

32
Heat Recovery and the Economizer Cycle

As discussed in Chapter 31, the heat rejected from cooling processes in hydronic heat pump systems often can be used to perform heating duty. In the case of hydronic heat pump systems, a common heat sink/heat source reservoir in the form of a water loop shared by a group of small heat pumps is used as the mechanism to transfer the rejected heat from heat pumps in cooling mode to heat pumps requiring heat. Figures 32.1 and 32.2 show some other forms of refrigeration-cycle heat recovery that utilize rejected heat from cooling for heating. Figure 32.1 shows a chiller with a double-bundle condenser to divert rejected heat as a part of the heating source. Figure 32.2 shows an air-cooled condenser used directly as a heating coil. These types of heat recovery systems, properly applied, can be effective ways to conserve thermal energy in an air-conditioning system.

The economizer cycle, as discussed in Chapter 5, is also an energy conservation scheme. An economizer cycle takes advantage of the cooler outside air in mild and cold weather to supplement or satisfy cooling needs. During mild weather, required cooling plant capacity can be reduced with an economizer cycle in operation. As the outside air temperature drops below the cold supply air temperature, the economizer cycle can satisfy the required supply air temperature, and the cooling plant can be shut down.

While both refrigeration cycle heat recovery and the economizer cycle are designed to conserve energy, the concepts of these two energy conservation schemes are quite different—to the point that they are basically incompatible. The economizer cycle saves cooling

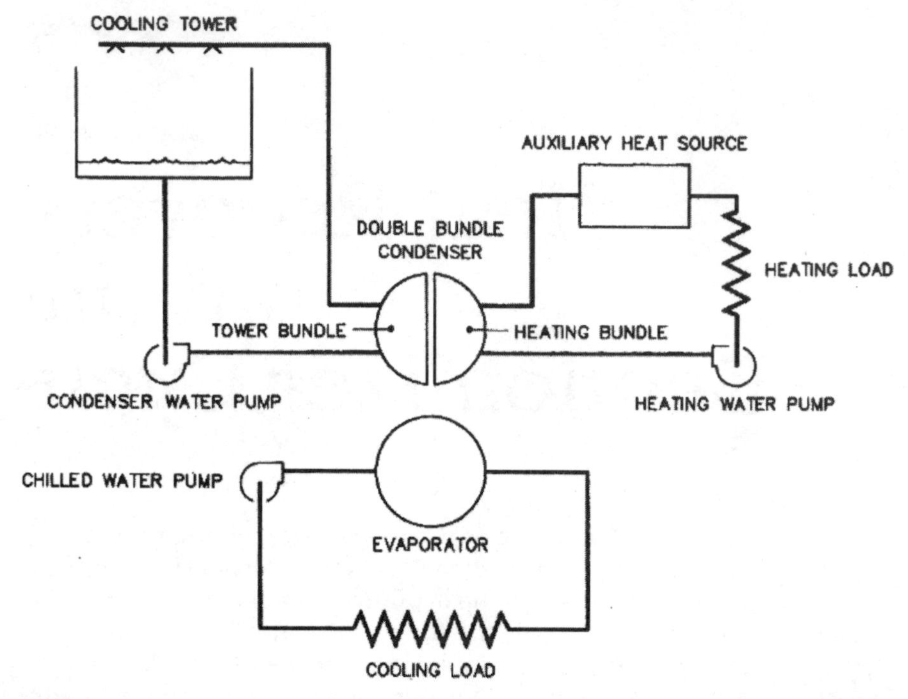

Figure 32.1. Double-bundle condenser.

energy. It is most effective when the refrigeration plant is shut down. On the other hand, the heat recovery scheme saves heating energy, and it relies on the operation of the refrigeration plant to transfer the rejected heat to satisfy the heating needs.

Purely from an energy conservation viewpoint, the refrigeration cycle heat recovery and the economizer cycle can always be used harmoniously together in a building air-conditioning system to obtain maximum operation economy. In such an application, the economizer cycle operates in conjunction with the heat recovery plant. Instead of shutting down the refrigeration plant completely to take full advantage of cool outside air, the economizer cycle can be controlled to maintain just enough refrigeration load for the plant to operate such that the rejected heat can satisfy the heating requirements.

Any one of three schemes—refrigeration-cycle heat recovery, an economizer cycle, or a combination heat recovery–economizer cycle—can be chosen as the air-conditioning system for a given building.

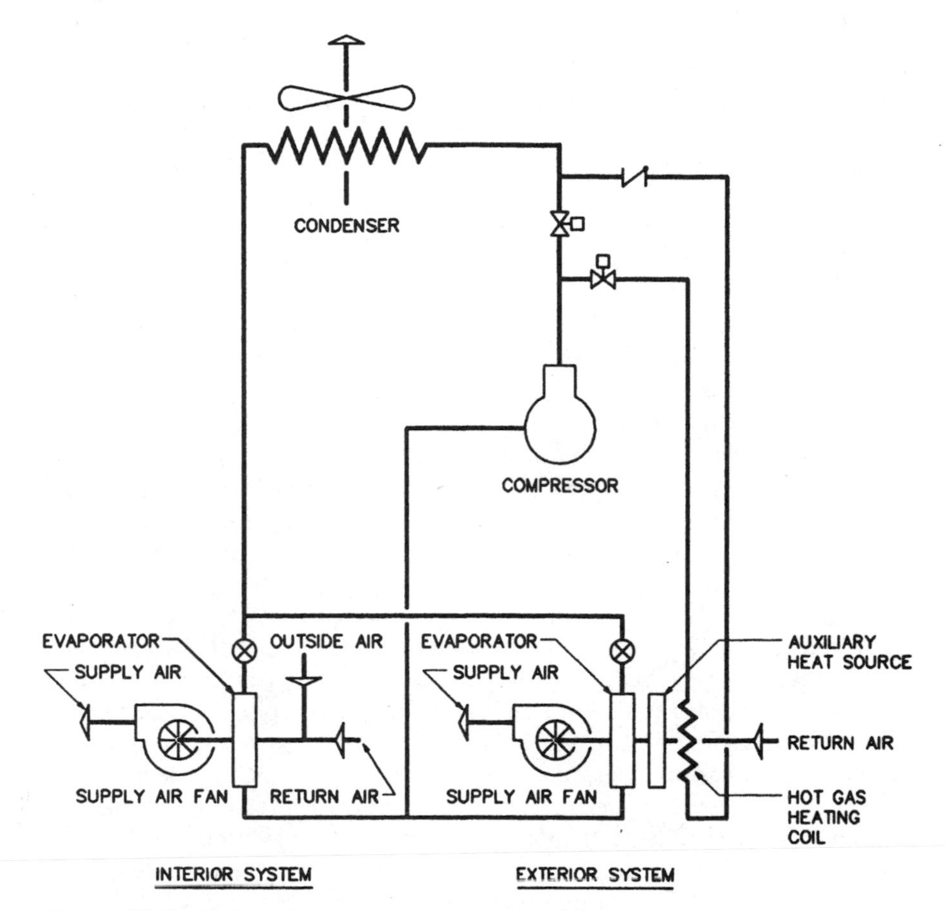

Figure 32.2. Hot gas heating.

Many important and interrelated factors, however, must be evaluated before an engineer can choose one of these design options. These factors include

1. Heating and cooling load profiles
2. Availability of energy sources and utility rates
3. Justification of initial investments
4. The choice of air handling systems
5. The requirements for humidity control

Generally, if the magnitude of the heating profile of a building is always minimal compared with concurrent cooling profiles, such as the load profiles of a well-insulated building with high internal loads, the choice will tend to favor the economizer cycle. This is so because there is minimal use of rejected heat to justify use of heat recovery, and much of the cooling energy can be saved with an economizer cycle. On the other hand, heat recovery will be a good choice if the building has steady and similar heating and cooling profiles for significant periods of time, since the rejected heat can always be used for heating that otherwise will have to be met by some form of new energy source. This is especially true for all-electric buildings where resistance heat is used as the heating source.

The availability of energy sources for a building may vary depending on the location and the type of building. Under different conditions, electricity, natural gas, propane, oil, and coal can all be considered as primary energy sources. Where two energy sources are available, the cost of each energy source can be an influential factor in the choice of systems.

Engineers are often asked to justify the additional expenditures required for energy conserving systems. In the case of the economizer cycle, these expenditures include the added return air fans, damper controls, and the possible increase in fan room space requirements. The additional expenditures for a heat recovery system include condenser modification, more controls, and larger pipes and deeper coils in the heating system because of temperature limitations in the heat recovery system. The costs of an auxiliary heating system and/or storage tanks which are often required for the heat recovery systems also must be included. A combination system involves expenditures for both systems mentioned above and for additional controls.

Detailed analysis must be done to evaluate the energy use of each scheme. The selection of air handling systems must be integrated into the evaluation process. While the use of heat recovery is practically independent of air handling system selection, use of an economizer cycle has a definite effect on air handling system selection. Energy savings of an economizer cycle used in conjunction with a single-duct system is straightforward. An economizer cycle applied to a double-duct system, however, will not perform as effectively as that applied to a single-duct system. When an economizer cycle is used with a conventional double-duct or multizone system, the heating profile is influenced by the system operation. When the economizer cycle is in operation in colder weather, it will always try to maintain the mixed air temperature to satisfy the cold air supply temperature. Heating is penalized because the heating coil will have to heat the air from the cold supply air temperature instead of the warmer mixed air temperature if the system did not incorporate an economizer cycle. For this reason, a heat recovery system is often a better energy conservation choice in a double-duct system.

The use of an economizer cycle does not penalize heating in a single-duct system. With a terminal reheat system, the reheat coil entering temperature is the same whether or not the system uses an economizer. With a variable volume dual-conduit system or variable volume system with perimeter heating, the heating system is completely independent of the cooling system, and an economizer cycle in the cooling system therefore has no effect on the heating system.

The humidification load in an air-conditioning system is primarily a function of the amount of outside air required. Since an economizer cycle relies on an increased amount of outside air to conserve thermal energy, the relative humidity in the conditioned space will decrease when the increased amount of cool outside air is drier than the humidity level to be maintained in the space. Figure 9.4 illustrates this effect for a double-duct system. Where an economizer cycle is used for conditioned spaces requiring high humidity levels, such as special computer rooms and surgical suites and nurseries in hospitals, it is possible that the cost of additional humidification may exceed the cooling cost savings. In this case, a heat recovery system could become the preferred energy conserving scheme. An article discussing the heat recovery versus economizer cycle issue in more detail is included in Appendix C.

33 Indirect Evaporative Cooling

Evaporative cooling is an adiabatic cooling process that cools air by evaporating water into an airstream. The latent heat of evaporation lowers the dry-bulb temperature of the air to produce the cooling sensation. This process is most effective in arid areas, where water evaporation potential is highest. No heat is removed from the airstream in an evaporative cooling process. There is no mechanical refrigeration energy savings in an air-conditioning system associated with an evaporative cooling process.

Indirect evaporative cooling is a process that utilizes the dry-bulb temperature depression of an evaporative cooling process to cool an airstream through an air-to-air heat exchanger. Indirect evaporative cooling is a sensible cooling process; i.e., sensible heat is removed from the airstream without increasing latent heat of the air. In general, the cooling effect obtained by the indirect evaporative cooling process is considered "free." When used in conjunction with mechanical refrigeration cooling, the indirect evaporative cooling process can save refrigeration energy. Figure 33.1 shows a schematic diagram of an indirect evaporative cooling system. Figure 33.2 shows an associated psychrometric plot of the process.

The commercially available indirect evaporative cooling system is often coupled with a direct evaporative cooling process to maximize depression of the dry-bulb temperature. The system is commonly referred to as an *indirect-direct evaporative cooling system.* In areas where the weather is hot and dry, this system not only lowers the dry-bulb temperature but also adds moisture to the airstream to provide better

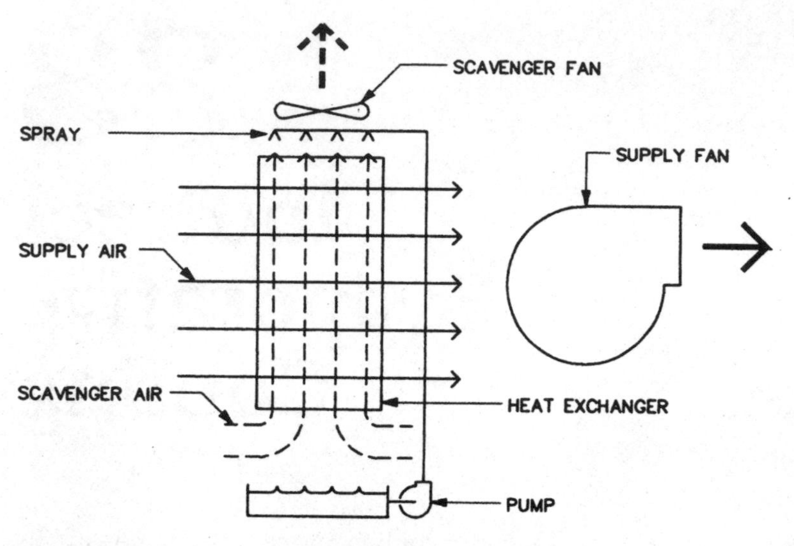

Figure 33.1. Indirect evaporative cooling unit.

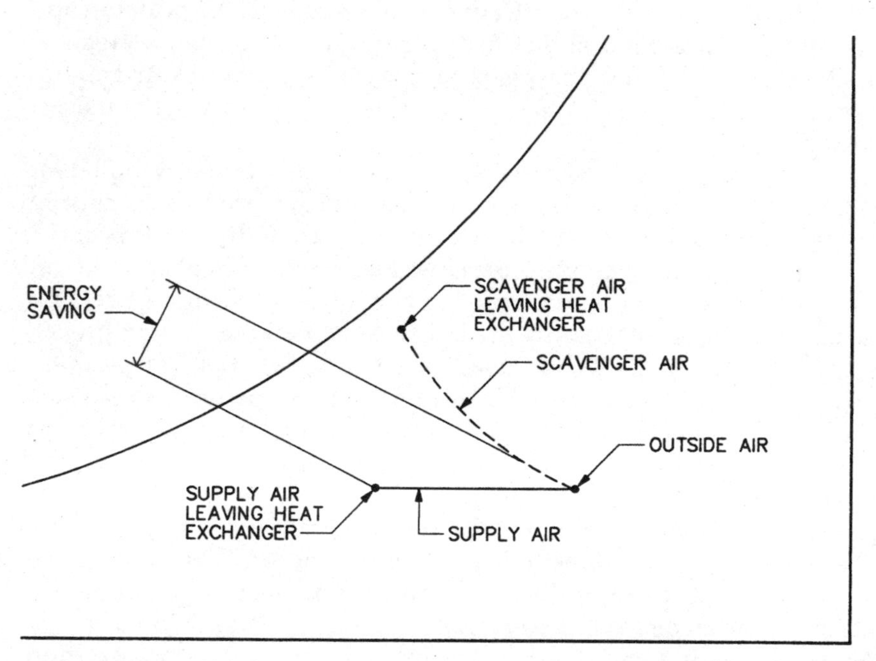

Figure 33.2. Psychrometric plot of an indirect evaporative cooling process.

comfort conditions. A typical psychrometric plot of an indirect-direct evaporative cooling process used in a space cooling application is shown in Figure 33.3. For installations where close temperature control is not essential, this system often replaces mechanical refrigeration. Where an indirect-direct evaporative cooling unit alone cannot achieve the required comfort condition or close temperature control is needed for the air-conditioning process, indirect or indirect-direct evaporative cooling units also can be used as precoolers in conjunction with mechanical refrigeration cooling to reduce installed mechanical cooling capacity and save refrigeration energy.

While the thermal energy savings from indirect evaporative cooling is considered free, fans and pumps of the indirect evaporative cooler require energy to operate, which must be taken into the overall considerations. The insertion of a precooler in the outside airstream will add static pressure to the air-conditioning system that will require more transportation energy to overcome. It should be noted that the added pressure drop through the cooler consumes transportation energy even when the precooler is not in operation during mild and cold

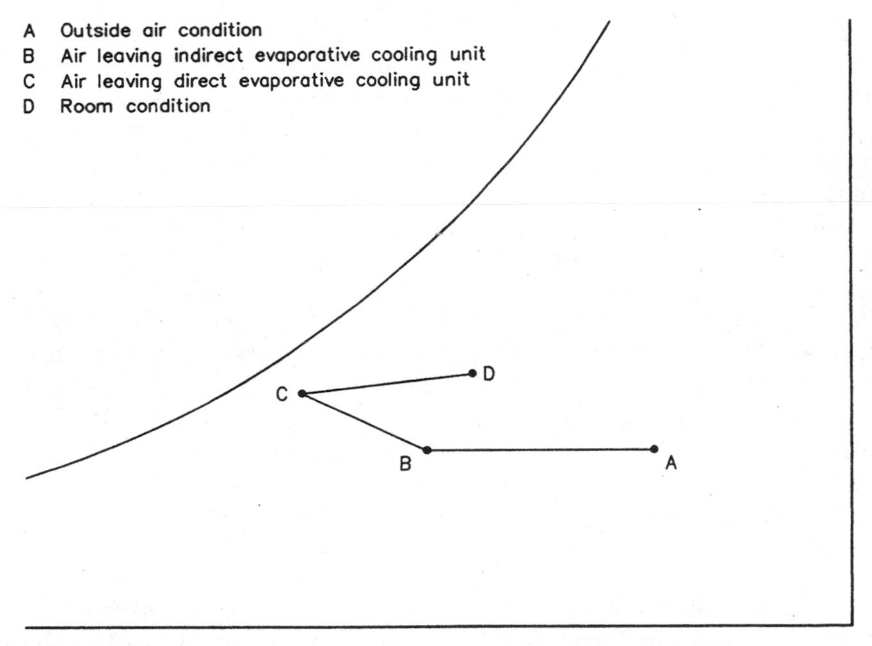

Figure 33.3. Psychrometric process of an indirect-direct evaporative cooling unit.

weather. These factors will diminish the thermal energy savings of the system. Evaluating the effectiveness of using indirect evaporative cooling to conserve thermal energy without taking these additional energy-consuming factors into consideration can produce grossly misleading results.

Indirect evaporative cooling is most effective when used in conjunction with 100 percent outside air systems. For systems with return air, the effectiveness of using an indirect evaporative cooling unit to precool the air drops considerably, since applying indirect evaporative cooling to return air is difficult to justify. When indirect evaporative cooling is applied to a system with return air, the pressure drop across the unit can be taken care of in two ways. An indirect evaporative cooler can be installed at the outside air intake, in which case the pressure drop across the cooler will be borne by the supply fan. For a system with return air, this added pressure drop across the outside air intake will have to be applied to the total supply air, which results in a bigger transportation energy penalty. Another way of taking care of this additional pressure drop is to install a blower in conjunction with the indirect evaporative cooling unit. With this arrangement, the system supply fan will not be burdened with the extra pressure drop across the indirect evaporative cooler. Where the cooler is used in conjunction with an air handling system using an economizer cycle, it also can be sized for 100 percent outside air even though the system is not a 100 percent outside air system. The larger cooler will help reduce energy consumption further during the period when the economizer is in operation.

To determine the effectiveness of an indirect evaporative cooling application, the amount of energy savings depends on what type of mechanical refrigeration plant is used. A large central chilled-water plant generally has an overall energy use of approximately 0.8 to 1 kW/ton. This kilowatt per ton figure includes power for the chillers, chilled-water pumps, condenser-water pumps, and cooling-tower fans. The comparable figure for a rooftop package air-conditioning unit, including power for refrigeration compressors and air-cooled condenser fans, on the other hand, can be as high as 1.3 to 1.5 kW/ton. Judging from this disparity, it is obvious that indirect evaporative coolers can be much more effective when used in conjunction with the rooftop packaged air-conditioning units.

The cost of electricity also plays an important role. In general, the higher the electricity cost, the easier it is to justify the use of indirect

	MILD CLIMATE WITH CENTRAL CHILLER PLANT	MILD CLIMATE W/ AIR COOLED ROOFTOP UNIT	ARID CLIMATE WITH CENTRAL CHILLER PLANT	ARID CLIMATE W/ AIR COOLED ROOFTOP UNIT
ENERGY SAVING, KWH	38,536	224,175	283,926	592,829
WATER UASGE, GALLON	NA	847,741	NA	1,403,457
EQUIPMENT COST, $	$175,000	$175,000	$175,000	$175,000
COST OF WATER, $/1000 GAL	NA	$1.10	NA	$1.10
TONNAGE SAVING COST, $	$89,000	$77,000	$109,000	$94,000
$/KWH	SIMPLE PAYBACK YEARS	SIMPLE PAYBACK YEARS	SIMPLE PAYBACK YEARS	SIMPLE PAYBACK YEARS
0.070	31.88	6.64	3.32	2.03
0.080	27.90	5.76	2.91	1.77
0.090	24.80	5.09	2.58	1.56
0.100	22.32	4.56	2.32	1.40
0.110	20.29	4.13	2.11	1.27
0.120	18.60	3.77	1.94	1.16

Figure 33.4. Impact on indirect evaporative cooling system of locale, type of air-conditioning system, and electricity costs.

evaporative cooling. Indirect evaporative cooling also uses a large amount of water, however. In the case of applying indirect evaporative cooling to air handling systems served from a central cooling plant, the cost of water does not become part of the evaluation because the cooling tower of the central plant uses the latent heat of water evaporation to remove the rejected heat, much the same as the water use in the indirect evaporative cooler. With rooftop units, however, water cost should become a part of the evaluation because the air-cooled condensing unit does not use water.

A small computer program can be written to simulate system operation with and without indirect evaporative cooling. Weather data such as monthly bin weather from *Engineering Weather Data,* published by the Department of Defense, which gives dry-bulb temperature bins with coincidental wet-bulb temperature, can be used as input. The spreadsheet shown in Figure 33.4 illustrates, as an example, the effect of the local climate, type of air-conditioning system, and cost of electricity, as discussed above. An article discussing a design experience with indirect evaporative cooling is included in Appendix D.

34

Energy Conservation and System Retrofit

Energy conservation is a term made popular in the mid-1970s, shortly after the Arab oil embargo. Prior to that time, many air-conditioning systems were designed to minimize construction costs and maximize architectural features of the building. Energy was plentiful in those days, and utility companies, instead of trying to conserve, were pushing to sell more energy. The emphasis in air-conditioning design shifted quickly in favor of energy conservation shortly after the oil crisis in the 1970s. There are, however, many buildings built before and even after the oil crisis that were designed with inefficient air handling systems that are still in operation today.

Energy conservation, by definition, is an integral part of the HVAC engineer's job. Today's engineer must be able to choose an air handling system for a given project that fits the needs of the building and at the same time finds a balance among lowest installation cost, minimum maintenance difficulties, and maximum energy efficiency.

From time to time, an engineer may be asked to retrofit an existing system for a variety of reasons. The building may need complete renovation, or a deteriorated HVAC system may have to be brought back to its original operating condition. Sometimes HVAC systems may have been identified as energy inefficient, and the building may need to be retrofitted with a more energy-efficient system. In this last case, construction money is spent to modify an existing air-conditioning system that is still operational to a more energy-conserving system.

Constant volume air handling systems generally use more thermal and transportation energy when compared with variable volume systems. There are many opportunities to modify existing constant volume systems to variable volume systems to conserve energy. Constant volume terminal reheat systems, as shown in Figure 8.1, can be converted to variable volume systems, as shown in Figure 24.1. The retrofit process involves converting constant volume terminals to variable volume terminals to minimize thermal energy waste and adding variable volume controls to the supply and return fans to conserve transportation energy.

The most effective and simplest way to modify a constant volume fan for variable volume application is to add a variable-frequency drive to the fan motor. Before variable-frequency drives became commercially available, other, mechanical modifications, such as adding inlet vanes, adding a movable disk at the inlet of the blower, or adding sliding cylinders over the fan wheel to effectively change the width of the wheel, had been used to vary the supply air volume.

Reheat coils can be disconnected in terminals that serve interior zones when a fixed minimum ventilation rate is not a requirement. Of course, for spaces requiring reheat for dehumidification control or to maintain minimum ventilation rates, existing constant volume terminals should not be changed to variable volume terminals.

Constant volume double-duct systems, as shown in Figure 9.1, also can be converted to variable volume applications, as shown in Figure 25.1. The retrofit process is basically the same as that for terminal reheat systems. Terminals have to be changed, and variable volume controls have to be added to the fans. Single-duct cooling-only terminals can be used for interior zones where a fixed minimum ventilation rate is not required.

Converting constant volume double-duct systems to variable volume operation can save thermal as well as transportation energy. Some conventional constant volume double-duct systems also can be modified to constant volume double-duct systems with subzone heating, as shown in Figure 10.1. This type of retrofit is particularly suitable for all-electric buildings, where an electric heater, instead of a hot-water coil, is used in the hot plenum. As discussed in Chapter 10, the subzone heating system uses less thermal energy compared with a conventional double-duct system. There will be more thermal energy savings with this retrofit compared with converting a constant volume double-duct system to a variable volume double-duct system. There will be no transportation energy savings with this type of retrofit, however, since the system stays as a constant volume system.

To retrofit a constant volume double-duct system to subzone heating, duct heaters need to be added to terminals serving exterior zones. Since the system stays as a constant volume system, no modification is needed to the terminals, and there will be no change at the air handling unit except the removal or simple deactivation of the heater in the hot plenum. Generally, the retrofit cost of this modification is considerably less than changing the system to variable volume operation.

To further conserve energy, a constant volume double-duct system also can be retrofitted to a variable volume subzone heating system, as shown in Figure 26.1. In this case, terminals will have to be changed, and variable volume controls will have to be added to the fans. While there is additional transportation energy to be saved with this retrofit, the worthiness of the additional retrofit cost required, however, is sometimes questionable.

Energy-conservation considerations cannot be stopped single-mindedly at saving energy alone. One must be reminded constantly that there is a cost associated with the comfort achieved by using air-conditioning systems. This cost should include both the installation costs and the ongoing energy costs, which must be balanced on sound economic principles. This is particularly true for retrofitting existing systems. The double-duct retrofit described above is a classic example. Many energy-conservation measures are accumulative. Most of the time, it is very easy to save a large amount of energy for a small investment at the beginning. As refined measures are added, the return on investment diminishes. There are cases where trying to save the last BTU can become a poor investment. Between these two extremes, there lies a wide spectrum of retrofitting possibilities that require careful engineering consideration. An article discussing an energy retrofit decision-making process is included in Appendix E.

Appendix A

Psychrometric Subroutine Uses ASHRAE Algorithms*

Easily Used Computer Subroutine Solves Universal Psychrometric Problems with Satisfactory Accuracy for Practical Applications

In the past few years, many engineers and computer programmers have been confronted with the task of solving psychrometric problems using computers. Algorithms for accurate psychrometric calculations were published by the National Bureau of Standards in January 1970,[1]† but unfortunately they are relatively lengthy and unsuitable for inclusion in engineering programs for practical applications where extreme accuracy is not mandatory.

Various simplified algorithms have been developed and programs written using different techniques to solve the psychrometric problems for particular applications.[2,3,4] Modifications to these existing programs are often required, and sometimes difficult to make, when one

*Reprinted with permission from *Heating/Piping/Air Conditioning,* October 1971.

†Superscript numerals indicate references at end of appendix.

tries to use them to fit a new program's needs. In this article, a standard compact FORTRAN subroutine is developed to solve "universal" psychrometric problems with a relatively high degree of accuracy for practical applications. The purpose for developing this subroutine is so that one can readily insert it into any main program in which common psychrometric properties have to be calculated.

Variables, Functions, Constants

The variables used in the subroutine are dry bulb temperature (DB) in degrees F, wet bulb temperature (WB) in degrees F, relative humidity (R) in percent, humidity ratio (W) in pounds of water per pound of dry air, enthalpy (H) in Btu per pound of dry air, dew point (DP) in degrees F, and an input index (M). The subroutine accepts as input a dry bulb temperature and any one of the remaining five psychrometric parameters, each corresponding to an index M. It calculates and passes back to the main program all of the associated variables.

Table 1 shows a listing of the subroutine, and Table 2 is a listing of a short testing program. Both are written in standard FORTRAN. Basically, the subroutine uses the algorithms developed by the ASHRAE Task Group for Energy Requirements, Heating and Cooling Load Calculation Subcommittee, chaired by Dr. T. Kusuda. Reference is also made to the 1967 *ASHRAE Handbook of Fundamentals*.[5]

The subroutine uses six interrelated functions, as follows:

1. PVSF calculates saturated vapor pressure at dry bulb temperature DB.
2. PV calculates vapor pressure at dry bulb temperature DB and wet bulb temperature WB.
3. WF calculates humidity ratio at dry bulb temperature DB and relative humidity R.
4. RHF calculates relative humidity at dry bulb temperature DB and humidity ratio W.
5. WBF calculates wet bulb temperature at dry bulb temperature DB and humidity ratio W.
6. DPF calculates dew point temperature at dry bulb temperature DB and humidity ratio W.

Table 1 Psychrometric Subroutine Using ASHRAE Algorithms

```
      SUBROUTINE PSYSUN(DB,WB,R,W,H,DP,M)
C---FOR M=1, INPUT=DB, WB      OUTPUT=R, W, H, DP
C---FOR M=2, INPUT=DB, R       OUTPUT=WB, W, H, DP
C---FOR M=3, INPUT=DB, W       OUTPUT=WB, R, H, DP
C---FOR M=4, INPUT=DB, H       OUTPUT=WB, R, W, DP
C---FOR M=5, INPUT=DB, DP      OUTPUT=WB, R, W, H
      DATA PB,FS/29.921,1.0045/
      GO TO(10,20,30,40,50),M
10    PVP=PV(DB,WB,PB,FS)
      W=0.622*FS*PVP/(PB-FS*PVP)
      R=PVP/PVSF(DB)
      GO TO 15
20    W=WF(DB,R,PB,FS)
      GO TO 25
50    PVP=PVSF(DP)
      W=0.622*FS*PVP/(PB-FS*PVP)
      GO TO 30
40    W=(H-0.24*DB)/(1061.+0.444*DB)
30    R=RHF(DB,W,PB,FS)
25    WB=WBF(DB,W,PB,FS)
      IF(M-5)15,45,15
15    DP=DPF(DB,W,PB,FS)
      IF(M-4)45,55,45
45    H=0.24*DB+(1061.+0.444*DB)*W
55    R=R*100.
      RETURN
      END
C---SATURATED VAPOR PRESSURE AT TEMPERATURE DB---------------
      FUNCTION PVSF(DB)
      DATA A,B,C/-7.90298,5.02808,-1.3816E-7/
      DATA D,E,F/11.344,8.1328E-3,-3.49149/
      DATA G,H,P,Q/-9.09718,-3.56654,0.876793,0.0060273/
      T=(DB+459.688)/1.8
      IF(T.LT.273.16) GO TO 3
      Z=373.16/T
      S=A*(Z-1.)+B*ALOG10(Z)+C*(10.**(D*(1-1./Z))-1.)
      S=S+E*(10.**(F*(Z-1.))-1.)
      GO TO 4
3     Z=273.16/T
      S=G*(Z-1.)+H*ALOG10(Z)+P*(1.-1./Z)+ALOG10(Q)
4     PVSF=29.921*10.**S
      RETURN
      END
C---VAPOR PRESSURE AT TEMPERATURES DB AND WB---------------
      FUNCTION PV(DB,WB,PB,FS)
      R=0.
      PVS=PVSF(WB)
      IF(DB.LE.WB) GO TO 4
      WS=0.622*PVS/(PB-PVS)
      IF(WB.GT.32.) GO TO 2
      PV=PVS-5.704E-4*PB*(DB-WB)/1.8
      GO TO 3
4     PV=PVS
      GO TO 3
2     CDB=(DB-32.)/1.8
      CWB=(WB-32.)/1.8
```

Table 1 Psychrometric Subroutine Using ASHRAE Algorithms (*Continued*)

```
      HL=597.31+C.4409*CDB-CWB
      CH=0.2402+0.4409*WS
      EX=(WS-CH*(CDB-CWB)/HL)/0.622
      PV=PB*EX/(1.+EX)
      IF(R.GT.0.) GO TO 3
      R=PV/PVSF(DB)
      IF(R.GT.0.1) GO TO 3
      WS=0.622*FS*PVS/(PB-FS*PVS)
      GO TO 2
 3    RETURN
      END
C---HUMIDITY RATIO AT TEMPERATURE DB AND RELATIVE HUMIDITY R
      FUNCTION WF(DB,R,PB,FS)
      PVS=PVSF(DB)
      WS=0.622*FS*PVS/(PB-FS*PVS)
      R=R*0.01
      DS=R*(PB-FS*PVS)/(PB-R*FS*PVS)
      WF=WS*DS
      RETURN
      END
C---RELATIVE HUMIDITY AT TEMPERATURE DB AND HUMIDITY RATIO W
      FUNCTION RHF(DB,W,PB,FS)
      PVS=PVSF(DB)
      WS=.622*FS*PVS/(PB-FS*PVS)
      DS=W/WS
      RHF=DS/(1.-(1.-DS)*FS*PVS/PB)
      RETURN
      END
C---WB TEMPERATURE AT TEMPERATURE DB AND HUMIDITY RATIO W
      FUNCTION WBF(DB,W,PB,FS)
      WBF=DB
      PVD=PB*W/((0.622+W)*FS)
 11   PVP=PV(DB,WBF,PB,FS)
      IF(PVP-PVD)20,30,10
 10   WBF=WBF-1.
      GO TO 11
 20   WBH=WBF+1.
      PVH=PV(DB,WBH,PB,FS)
      X=(PVD-PVP)/(PVH-PVP)
      WBF=WBH*X+WBF*(1.-X)
 30   RETURN
      END
C---DP TEMPERATURE AT TEMPERATURE DB AND HUMIDITY RATIO W
      FUNCTION DPF(DB,W,PB,FS)
      DPF=DB
      PVD=PB*W/((0.622+W)*FS)
 11   PVS=PVSF(DPF)
      IF(PVS-PVD)20,30,10
 10   DPF=DPF-1.
      GO TO 11
 20   DPH=DPF+1.
      PVH=PVSF(DPH)
      X=(PVD-PVS)/(PVH-PVS)
      DPF=DPH*X+DPF*(1.-X)
 30   RETURN
      END
```

Table 2 Test Program for Psychrometric Subroutine

```
      READ 2, DB,X,M
      GO TO (10,20,30,40,50),M
10    WB=X
      GO TO 60
20    R=X
      GO TO 60
30    W=X
      GO TO 60
40    H=X
      GO TO 60
50    DP=X
60    CALL PSYSUN(DB,WB,R,W,H,DP,M)
      PRINT 1,M,DB,WB,R,W,H,DP
1     FORMAT(2X,I2,3F6.1,2F10.5,F6.1//)
2     FORMAT(2F8.0,I1)
      STOP
      END
```

Two constants are used throughout the subroutine: the barometric pressure (PB), which is set at 29.921 in. Hg, and a vapor pressure correction factor (FS), which is set at 1.0045. The exact value of FS is a function of PB and DB, and tabulated values of FS can be found in Table 5 of Reference 5. In the subroutine, FS is made a constant rather than a function of PB and DB for simplification; a relatively high degree of accuracy is retained despite this.

How Calculations Are Handled

Function PVSF is used in all of the associated functions in the subroutine. The algorithms of this function and function PV are identical to those developed in Reference 2 with the exception of introducing correction factor FS to the function PV for relatively dry air (R less than 10 percent).

Function WF uses function PVSF and the following equations:[5]

$$WS = 0.622*FS*PVS/(PB - FS*PVS) \tag{1}$$

$$R = DS/(1. - (1. - DS)*FS*PVS/PB) \tag{2}$$

Rearranging Equation 2, we get:

$$DS = R^*(PB - FS^*PVS)/(PB - R^*FS^*PVS) \tag{3}$$

$$W = WS^*DS \tag{4}$$

where WS = humidity ratio of saturated air at dry bulb temperature DB
PVS = saturated vapor pressure at dry bulb temperature DB
DS = degree of saturation

Function RHF uses function PVSF and Equations 1 and 2, where DS in Equation 2 is obtained by rearranging Equation 4.

Function WBF uses an iteration method. Vapor pressure (PVD) of the given condition is first calculated by the equation:

$$PVD = PB^*W/((0.622 + W)^*FS) \tag{5}$$

The iteration process starts by assuming WB equal to DB. Vapor pressure of this trial WB-DB condition is calculated using function PV and compared with PVD. WB is then decreased at 1 F intervals until the trial vapor pressure is less than PVD. The final WB value is calculated by assuming that the vapor pressure variation within a 1 F range is linear, and straight line interpolation is applied. It should be noted that the wet bulb temperature thus obtained is an approximate value. It is quite adequate for use in evaluating air conditioning processes, however.

Function DPF also uses an iteration method. The only difference between functions DPF and WBF is that the former uses DP instead of WB and function PVSF instead of function PV.

The volume of moist air is not calculated in the subroutine since it is not considered as a common psychrometric property dealt with in air conditioning processes. It can easily be added if desired, however, by using the following equation:[2]

$$V = 0.754^*(DB + 459.7)^*(1. + 7000.^*W/4360.)/PB \tag{6}$$

This is solved after W is either given to or calculated by the subroutine.

For the high altitude regions such as Colorado, New Mexico, and Utah, psychrometric calculations for sea level conditions are not acceptable. An altitude change can readily be accomplished in the subroutine by either reassigning values of PB and FS in the DATA statement in the subroutine or making PB and FS part of the arguments of the subroutine. For the latter case, the values of PB and FS can be made part of the main program input if the program is written to consider altitude as one of the design parameters.

Some Sample Results

Tables 3, 4, and 5 give sample results calculated using this subroutine (Columns B) and the corresponding accurate values given in Reference

Table 3 Sample Results Using Psychrometric Subroutine (Columns B) Are Compared with Correct Results from Reference 1 (Columns A) for Various Dry Bulb and Wet Bulb Conditions at Sea Level with DB and WB Used as Input (PB = 29.921 and FS = 1.0045).

		R		W		H		DP	
DB	WB	A	B	A	B	A	B	A	B
0.	−2.	39.9	39.6	0.00031	0.00031	0.33	0.33	−16.9	−17.0
40.	30.	28.5	28.1	0.00148	0.00146	11.20	11.17	12.3	12.1
80.	67.	51.1	51.1	0.01122	0.01121	31.51	31.49	60.3	60.3
80.	50.	4.1	4.1	0.00088	0.00089	20.18	20.17	2.1	2.3
120.	100.	50.0	50.0	0.03822	0.03819	71.39	71.36	96.2	96.2
120.	70.	6.0	6.0	0.00432	0.00436	33.65	33.66	35.3	35.5
160.	140.	59.0	59.0	0.14730	0.14707	205.02	204.89	138.7	138.7
160.	80.	1.9	1.9	0.00383	0.00389	42.80	42.80	32.2	32.6

Table 4 Sample Results Using Psychrometric Subroutine (Columns B) Are Compared with Correct Results from Reference 1 (Columns A) for Various Dry Bulb and Wet Bulb Conditions at 5000 ft with DB and WB Used as Input (PB = 24.89 and FS = 1.0040).

		R		W		H		DP	
DB	WB	A	B	A	B	A	B	A	B
0.	−2.	48.4	48.1	0.00046	0.00045	0.51	0.48	−13.5	−13.6
40.	30.	34.9	34.6	0.00217	0.00216	11.97	11.93	16.4	16.2
80.	67.	53.4	53.4	0.01415	0.01415	34.74	34.71	61.6	61.6
80.	50.	9.3	9.4	0.00242	0.00244	21.89	21.88	18.7	18.8
120.	100.	51.1	51.1	0.04757	0.04755	81.83	81.79	96.9	96.9
120.	70.	8.6	8.7	0.00754	0.00759	37.25	37.26	44.7	44.9
160.	140.	59.4	59.4	0.18744	0.18718	250.43	250.30	139.0	139.0
160.	80.	3.4	3.4	0.00837	0.00845	47.95	47.96	47.7	47.7

Table 5 Sample Results Using Psychrometric Subroutine (Columns B) Are Compared with Correct Results from Reference 1 (Columns A) for Various Dry Bulb and Wet Bulb Conditions at Sea Level with DB and R Used as Input (PB = 29.921 and FS = 1.0045).

		WB		W		H		DP	
DB	R	A	B	A	B	A	B	A	B
0.	39.9	-2.	-2.0	0.00031	0.00031	0.33	0.33	-16.9	-16.9
40.	28.5	30.	30.1	0.00148	0.00148	11.20	11.19	12.3	12.3
80.	51.1	67.	67.0	0.01122	0.01121	31.51	31.50	60.3	60.3
80.	4.1	50.	50.0	0.00088	0.00089	20.18	20.17	2.1	2.2
120.	50.0	100.	100.0	0.03822	0.03821	71.39	71.37	96.2	96.2
120.	6.0	70.	70.0	0.00432	0.00435	33.65	33.65	35.3	35.4
160.	59.0	140.	140.0	0.14730	0.14709	205.02	204.91	138.7	138.7
160.	1.9	80.	80.0	0.00383	0.00385	42.80	42.76	32.2	32.4

1 (Columns A). Tables 3 and 4 use DB and WB as input whereas Table 5 uses DB and R as input. Tables 3 and 5 are for sea level conditions, and Table 4 is for an altitude of 5000 ft. Note that the subroutine gives very good results throughout the wide DB-WB spectrum. Test results using index M values other than 1 and 2 yield the same degree of accuracy.

The functional approach of this subroutine makes it simple for the user to delete functions not required in his particular program. The related functions can also be used independently to calculate single variables.

A portion of this subroutine has been used in the new version of the Heating and Cooling Load Calculation Program (HCC III) recently developed by Automated Procedures for Engineering Consultants, Inc. A subsequent article will discuss psychrometric analysis of air conditioning systems by computer.

References

1. Kusuda, T., *Algorithms for Psychrometric Calculations (Skeleton Tables for the Thermodynamic Properties of Moist Air)*, Building Science Series 21, National Bureau of Standards, January 1970.
2. *Algorithms for Building Heat Transfer Subroutines*, ASHRAE Task Group for Energy Requirements, April 1968.

3. Fairfax, Tung, and Courtney, "Further Investigations of Empirical Equations Used in Programming a Psychrometric Table on a Computer," *ASHRAE Journal,* January 1969.
4. Traylor and Suggs, "Apparatus Dew Point Determined by Computer," *ASHRAE Journal,* January 1969.
5. *ASHRAE Handbook of Fundamentals,* Chapter 6, "Psychrometrics," American Society of Heating, Refrigerating and Air-Conditioning Engineers, Inc., 1967.

Appendix **B**

Psychrometric Analysis of AC Systems by Computer*

Computer Program for Psychrometric Analysis of Air Conditioning Systems Offers User the Options of Fixing Supply Air Quantity or Providing Space Humidity Control by Varying Supply Air Volume or Providing Reheat

Once the psychrometric properties of moist air at any given state can be determined by a computer program,[1]† the psychrometric analysis of an air conditioning system by computer becomes a relatively easy task. The results of such an analysis are virtually unlimited. This article discusses a system analysis program, SYSSUN, that can readily be used for the system load calculation. The program, written in standard FORTRAN, is listed in Table 1. It uses the psychrometric subroutine PSYSUN and its interrelated functions, which were discussed in an article last month.[1]

*Reprinted with permission from *Heating/Piping/Air Conditioning*, October 1971.
†Superscript numerals indicate references at end of appendix.

Table 1 FORTRAN Listing of System Analysis Program SYSSUN

```
C-A SYSTEM ANALYSIS PROGRAM CALCULATING SPACE AND COIL CONDI-
C-TIONS C-W/ OR W/O SPACE HUMIDITY CONTROL ..................
      DATA PB, FS/29.921, 1.0045/
      READ 1,TZ,RZ,TO,WBO,OSA,ZS,ZL,ZR,DTR,BF,SP,LSF,IADP
1     FORMAT (4F3.0,F5.0,3F7.0,F3.0,F4.3,F3.0,2I1)
      WZ=WF (TZ,RZ,PB,FS)
      M=1
      CALL PSYSUN (TO,WBO,RO,WO,HO,DPO,M)
      KGO=0
      WL=0.
C-CALCULATE FAN HEAT DISTRIBUTION AND CFM ....................
      DTF=0.364*SP
      GO TO (5,10), LSF
C-DRAW THRU..................................................
5     ZDTF=DTF
      ZTFR=0.
      GO TO 15
C-BLOW THRU..................................................
10    ZDTF=DTF*0.65
      ZTFR=DTF-ZDTF
15    TL1=TZ-DTR-ZDTF
      SCFM=ZS/(1.1*DTR)
C-CALCULATE MIXED AIR CONDITION..............................
20    TR=TZ+ZR/(1.1*SCFM)
      PO=OSA/SCFM
      TM=TO*PO+TR*(1.-PO)
C-CALCULATE VARIOUS SENSIBLE HEAT............................
      OASH=OSA*1.1*(TO-TR)
      SFHT=SCFM*SP*0.4
      GS=ZS+ZR+OASH+SFHT
      ZSF=ZS+ZDTF*1.1*SCFM
      TMA=TM+ZTFR
      ESH=ZSF+BF*(ZTFR*1.1*SCFM+OASH+ZR)
C-CALCULATE VARIOUS LATENT HEAT..............................
25    OALH=OSA*4840.*(WO-WZ)
      WMA=WO*PO+WZ*(1.-PO)
      ELH=ZL+BF*OALH
      GL=ZL+OALH
C-CALCULATE VARIOUS SENSIBLE HEAT FACTORS ...................
      ESHF=ESH/(ESH+ELH)
      GSHF=GS/(GS+GL)
      ZSHF=ZSF/(ZSF+ZL)
      KA=1
      GO TO (30,35), IADP
C-IADP=1. NO SPACE HUMIDITY CONTROL, CFM FIXED BY INPUT DTR
30    IF(KGO) 40, 40, 45
40    TADP=(TL1-BF*TMA)/(1.-BF)
      WADP=WF(TADP, 100., PB, FS)
      WZ1=WZ
45    WZ=WADP+(TZ-TADP)*(1./ESHF-1.)/4410.
      DWZ=(WZ-WZ1)*10000000.
      IF(ABS(DWZ).GE.1.) GO TO 25
      TL=TL1
      WL=WMA-(TMA-TL)*(1./GSHF-1.)/4410.
      GO TO 50
C-IADP=2. SPACE HUMIDITY CONTROLLED USING ADP METHOD W/
C-REHEAT ....................................................
35    TADP=DPF(TZ,WZ,PB,FS)
55    WADP=WF(TADP,100.,PB,FS)
      TESHF=1./(1.+4410.*(WZ-WADP)/(TZ-TADP))
```

Table 1 FORTRAN Listing of System Analysis Program SYSSUN (*Continued*)

```
      GO TO (60,65),KA
60    IF(TESHF-ESHF) 70, 75, 80
80    TADP=TADP-1.
      IF(TADP-32.)85, 85, 55
70    ESHFL=TESHF
      TADPL=TADP
      TADP=TADP+1.
      KA=2
      GO TO 55
65    X=(ESHF-ESHFL)/(TESHF-ESHFL)
      TADP=TADP*X+TADPL*(1.-X)
75    TL2=TADP+BF*(TMA-TADP)
      TL=TL2
      IF(KGO.GT.0.) GO TO 50
      IF(TL2-TL1) 85, 100, 100
C-REHEAT REQUIRED ........................................
85    WL=WZ-(TZ-TL1)*(1./ZSHF-1.)/4410.
      WADP=(WL-BF*WMA)/(1.-BF)
      TADP=DPF(TL1,WADP,PB,FS)
      TL2=TADP+BF*(TMA-TADP)
      RHEAT=SCFM*1.1*(TL1-TL2)
      RHLT=TL1
      TL=TL2
      GO TO 105
C-REHEAT NOT REQUIRED ....................................
100   WL1=WL
      SCFM=ZSF/(1.1*(TZ-TL2))
      WL=WMA-(TMA-TL)*(1./GSHF-1.)/4410.
      IF(WL1.LE.0.) GO TO 20
      DWL=(WL-WL1)*10000000.
      IF(ABS(DWL)-1.) 50,20,20
50    RHEAT=0.
      RHLT=TL
105   M=3
C-CALCULATE VARIOUS PSYCHROMETRIC PROPERTIES AND COIL LOAD
      CALL PSYSUN (TMA,WBE,RE,WMA,HE,DPE,M)
      CALL PSYSUN (TL,WBL,RL,WL,HL,DPL,M)
      CALL PSYSUN (TZ, WBZ RZ, WZ HZ, DPZ, M)
      Q=SCFM*4.5*(HE-HL)
      WBM=WBF(TM WMA,PB,FS)
      RM=RHF(TM,WMA PB,FS)*100
C-PRINT OUT RESULTS ......................................
      PRINT 115
      FORMAT (//19X"DB"4X"WB"4X"RH"6X"W"7X"H"5X"DP 4X ADP"/)
      PRINT 90 TMA,WBE,RE,WMA HE DPE,TL,WBL,RL,WL,HL,DPL,TADP
90    FORMAT (3X"COIL ENT "4X3F6 1,F9.5,F7.2,F6.1//
      3X"COIL LVG."4X.3F6.1,F9.5,F7.2,2F6.1/)
      PRINT 95 RHLT,TZ,WBZ,RZ,WZ,HZ,DPZ,TO,WBO,RO,WO,HO,DPO,TR,
     -TM,WBM,RM
95    FORMAT (3X"REHEAT LVG."F8.1//3X"ZONE COND."3X,3F6.1,F9.5,
     -F7.2,F6.1//3X"OSA COND."4X3F6.1,F9.5,F7.2,F6.1//3X"RETURN"
     -1X"AIR"3XF6.1//3X"MIXED AIR"4X,3F6.1/)
      PRINT 110,SCFM,Q,RHEAT
110   FORMAT (3X"SUPPLY CFM"9X"="F10.0//
     -3X"COOLING COIL LOAD ="F10.0//3X"REHEAT COIL LOAD"
     -3X"="F10.0)
      STOP
      END
```

Two Approaches to the Problem

The program provides two independent approaches, selected by the user. One approach is to fix the supply air quantity by presetting the supply air temperature to the conditioning space. The space humidity condition in this case is dictated by coil performance. For a given coil, set by a given bypass factor,[2] the humidity in the conditioned space may rise to higher than "normal" for a space with a low sensible heat factor, or it may drop to lower than "normal" for a space with a high sensible heat factor—*normal* being used here to designate the optimum design value.

The second approach provides the space humidity control by either varying the supply air quantity, accomplished by adjusting the supply air temperature to the conditioned space, or reheating the supply air when the supply air temperature drops below a preset limit.

Fig. 1 shows the input form of the system analysis program. Input variables, together with other variables used in the program, are listed in Table 2.

Fig. 2 illustrates a typical psychrometric process for a blow-through air conditioning system. Line 1-2 represents the temperature rise due to heat retained in return air. Line 4-5 represents the heat generated at the supply fan. The amount of heat generated by the supply fan, in Btuh, is calculated as:

$$\begin{aligned} \mathrm{SHFT} &= 0.0001573{*}2545{*}\mathrm{SP}{*}\mathrm{SCFM} \\ &= 0.4{*}\mathrm{SP}{*}\mathrm{SCFM} \qquad (1) \end{aligned}$$

The corresponding temperature rise is:

$$\begin{aligned} \mathrm{DTF} &= \mathrm{SFHT}/(1.1{*}\mathrm{SCFM}) \\ &= 0.364{*}\mathrm{SP} \qquad (2) \end{aligned}$$

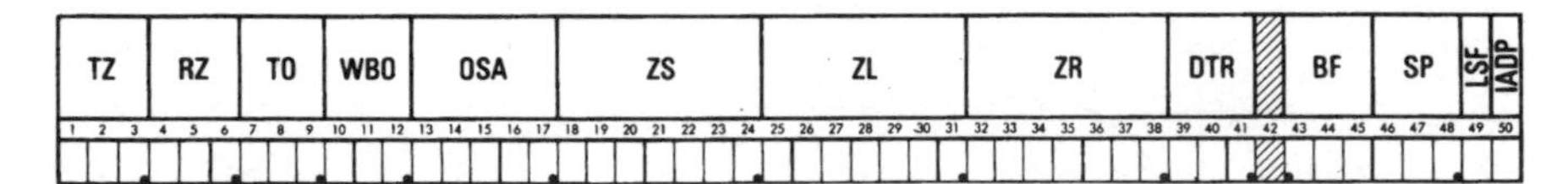

Figure 1. Input form of analysis program SYSSUN.

Table 2 List of Variables Used in System Analysis Program SYSSUN

A. Dry bulb temperatures, F:
TO = outdoor air
TMA = mixed air after fan (coil entering)
TM = mixed air before fan
TR = return air
TZ = conditioned space air
TL = cooling coil leaving air
TL1 = coil leaving air with fixed cfm
TL2 = coil leaving air with humidity control
TADP = apparatus dew point
RHLT = reheat coil leaving air
B. Wet bulb temperatures, F:
WBO = outdoor air
WBM = mixed air before fan
WBZ = conditioned space air
WBE = coil entering air (mixed air after fan)
WBL = coil leaving air
C. Humidity ratios, lb water vapor per lb dry air:
WO = outdoor air
WMA = mixed air (coil entering)
WZ = conditioned space air
WL = coil leaving air
WADP = apparatus dew point
D. Relative humidities, percent:
RO = outdoor air
RE = coil entering air (mixed air after fan)
RM = mixed air before fan
RZ = conditioned space air
RL = coil leaving air
E. Enthalpies, Btu per lb dry air:
HO = outdoor air
HE = coil entering air
HZ = conditioned space air
HL = coil leaving air
F. Dew Point temperatures, F:
DPO = outdoor air
DPE = coil entering air
DPZ = conditioned space air
DPL = coil leaving air

Table 2 List of Variables Used in System Analysis Program SYSSUN (*Continued*)

G. Various loads, Btuh:
ZS = zone sensible
ZL = zone latent
ZR = zone return air
ZSF = zone sensible plus fan heat downstream from cooling coil
SFHT = fan heat, total
OASH = outdoor air sensible
OASL = outdoor air latent
GS = grand total sensible
GL = grand total latent
ESH = effective sensible
ELH = effective latent
Q = cooling coil
RHEAT = reheat coil
H. Sensible heat factors, decimal:
ESHF = effective
GSHF = grand
ZSHF = zone
TESHF = trial effective
I. Air quantities, cfm:
OSA = outdoor air
SCFM = supply air
J. Temperature differentials, F:
DTR = between conditioned space and supply air
DTF = rise due to fan heat
ZDTF = rise due to fan heat downstream from cooling coil
ZTFR = rise due to fan heat at fan
K. Other variables:
PB = barometric pressure, in. Hg
FS = vapor pressure correction factor
BF = coil bypass factor
SP = fan static pressure, in, WG
LSF = supply fan location
1 = draw-through
2 = blow-through
IADP = calculation method
1 = fixed supply air
2 = humidity control
PO = outdoor air, percent
M, KGO, KA = operating indexes

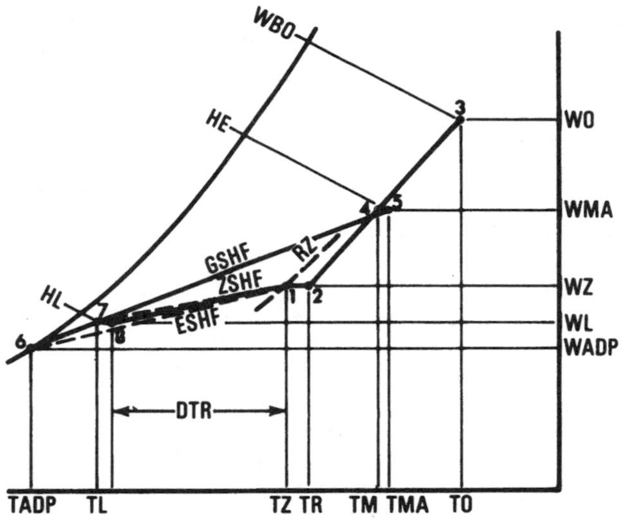

Figure 2. Psychrometric plot for a blow-through system without reheat.

It should be noted that not all of the fan heat can be converted into temperature rise at the location of the fan. Part of the energy will be dissipated in various parts of the system. This program assumes that 35 percent of the fan heat is dissipated at the fan (Line 4-5) and 65 percent of the fan heat is dissipated along the supply path downstream from the air conditioning apparatus (Line 7-8).

The effective sensible heat factor, ESHF, which is the ratio of effective zone sensible heat to the effective zone sensible and latent heat, has been defined in Reference 2 (page 1-122). Effective sensible heat, ESH, as defined therein, however, should be expanded to include heat to return air and fan heat. The new definition of ESH for a draw-through system is: zone sensible heat plus fan heat plus that portion of the outdoor air sensible heat and return air heat which is considered as being bypassed, unaltered, through the air conditioning apparatus. The definition of ESH for a blow-through system is: zone sensible heat plus a portion of the fan heat considered as being dissipated along the supply path downstream from the air conditioning apparatus plus that portion of the outdoor air sensible heat, return air heat, and the remaining portion of the fan heat which is considered as being bypassed, unaltered, through the apparatus.

The program first uses the psychrometric subroutine and individual psychrometric functions to calculate all psychrometric properties of the outdoor air and the humidity ratio of the conditioned space, WZ. It calculates next the required air temperature leaving the cooling coil, TL1, and the supply air quantity, SCFM, based on ZDTR and the calculated DTF. The air quantity thus calculated is considered fixed and final for IADP = 1, but it is considered as a trial value for IADP = 2. The program then proceeds to calculate the return air temperature; the mixed condition of return and outdoor air, TM and WMA; the coil entering condition, TMA and WMA; the zone sensible heat factor, ZSHF; the effective sensible heat factor, ESHF; and the grand sensible heat factor, GSHF.

Case 1: Fixed Supply Air Quantity

For the first approach, IADP = 1, the program then calculates the apparatus dew point, TADP, which is a function of TMA, TL1, and BF, bypass factor:

$$\text{TL1} = \text{TADP} + \text{BF*(TMA} - \text{TADP)} \tag{3}$$

$$\text{TADP} = (\text{TL1} - \text{BF*TMA})/(1 - \text{BF}) \tag{4}$$

The humidity ratio of TADP at saturation, WADP, is a function of TADP and relative humidity, which in this case is equal to 100 percent (Function WF, Reference 1). With TADP and WADP known, the humidity ratio maintained in the conditioned space can be calculated from the equation:

$$\text{WZ} = \text{WADP} + (\text{TZ} - \text{TADP})\text{*}(1/\text{ESHF} - 1)/4410 \tag{5}$$

The more familiar form of this equation, used to calculate sensible heat factor—in this case, ESHF—between two given points on a psychrometric chart, is:

$$\text{ESHF} = 0.244\text{*}(\text{TZ} - \text{TADP})/(0.244\text{*}(\text{TZ} - \text{TADP}) + 1076\text{*}(\text{WZ} - \text{WADP}))$$

$$= 1/(1 + 4410\text{*}(\text{WZ} - \text{WADP})/(\text{TZ} - \text{TADP})) \tag{6}$$

The calculated value of WZ will differ from the value of WZ set by the input conditions because of the imposed limitation of the coil. Using the calculated WZ, the computer recalculates all latent loads, sensible heat factors, and a new value of WZ. This iteration process continues until the difference between two successive values of WZ is less than 1×10^{-7}. The humidity ratio of the supply air, WL, is calculated using Equation 6 in a different form:

$$WL = WMA - (TMA - TL)*(1/GSHF - 1)/4410 \qquad (7)$$

The coil entering and leaving enthalpies, HE and HL, together with other psychrometric properties, are calculated next using Subroutine PSYSUN. Finally, the coil load is calculated as:

$$Q = SCFM*4.5*(HE - HL) \qquad (8)$$

Case 2: Controlled Space Humidity

For the second approach, IADP = 2, to maintain WZ, the program must find the TADP that together with WZ will yield the calculated ESHF. An iteration method is used by assuming a starting trial TADP equal to the dew point temperature of the conditioned space, which is a function of TZ and WZ (Function DPF, Reference 1). The trial WADP corresponding to TADP at saturation is calculated next using Function WF. The program then decreases TADP at 1° F intervals and calculates the corresponding ESHF using Equation 6 until the trial ESHF is less than the ESHF of the conditioned space. The final TADP is found by assuming that the ESHF variation within a 1°F range of TADP is linear and applying straight line interpolation. This program assumes that when TADP drops below 32°F, the cooling coil selection becomes impractical. A cooling coil with higher TADP will be used with reheat.

Humidity Control by Increased Air Volume

With TADP known, the supply air temperature required, TL2, can be calculated using Equation 3 with TL1 replaced by TL2. Comparing TL2 with previously calculated TL1, for TL2 higher than TL1, the program iterates by recalculating SCFM, return and mixed air temperatures, sensible heat factors, and WL. The iteration process continues until the difference between two successive values of WL is less than 1×10^{-7}. The coil entering and leaving conditions and the coil load are then calculated in a manner similar to that used in Case 1.

Humidity Control by Reheat

Reheat is needed if TL2 is lower than TL1 since TL1 is the low limit of the supply air temperature. Fig. 3 shows the psychrometric process for this case. The program calculates the humidity ratio corresponding to TL1, which in this case is the supply air temperature leaving the reheat coil, RHLT:

$$WL = WZ - (TZ - TL1)*(1/ZSHF - 1)/4410 \quad (9)$$

Since WL is the same for air leaving cooling coil and reheat coil, WADP for the cooling coil can be calculated as:

$$WADP = (WL - BF*WMA)/(1 - BF) \quad (10)$$

A new value of TADP for the cooling coil with a reheat condition corresponding to WADP can be calculated using Function DPF. Note that a dry bulb temperature higher than TADP is needed to start the

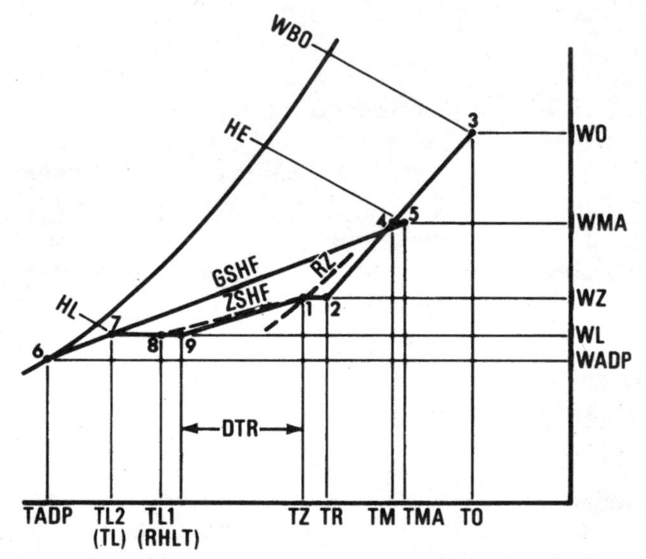

Figure 3. Psychrometric plot for a blow-through system with reheat.

iteration process in Function DFP. TL1 is chosen in this case since it is the known temperature closest to TADP. The value of TL2 corresponding to the new TADP is again calculated using Equation 3. Finally, the program proceeds to calculate reheat and cooling coil loads.

Some Examples

Four examples are presented in Figs. 4 through 7 to illustrate various results of the program. The first, in Fig. 4, illustrates a low latent load condition (ZSHF = 1) without space humidity control; the second, in Fig. 5, illustrates the low latent load condition with humidity control; the third, in Fig. 6, illustrates a high latent load condition (ZSHF = 0.667) without space humidity control; and the fourth, in Fig. 7, illustrates the high latent load condition with humidity control.

Note the relative humidity variations in the conditioned space for Case 1, Figs. 4 and 6, the increased air quantity for humidity control when latent load is low, Fig. 5; and the need for reheat for humidity control when latent load is high, Fig. 7.

It should be noted that outputs shown in Figs. 4 through 7 deviate from the formats programmed in SYSSUN for added clarity.

TZ	RZ	TO	WBO	OSA	ZS	ZL	ZR	DTR		BF	SP	LSF	IADP
75.	50.	95.	75.	200.	22000.	.	4000.	20.		.1	4.	2	1

	DB	WB	RH	W	H	DP	ADP
Coil entering	82.4	65.1	39.2	0.00927	29.96	55.1	
Coil leaving	54.1	52.5	90.6	0.00806	21.72	51.4	50.9
Reheat leaving	54.1						
Zone condition	75.0	60.7	43.5	0.00806	26.82	51.3	
Outside air condition	95.0	75.0	39.8	0.01412	38.37	66.7	
Return air	78.6						
Mixed air	81.9	64.9	39.8				

Supply volume, cfm = 1000
Cooling coil load, Btuh = 37,060
Reheat coil load, Btuh = 0

Figure 4. Program input and results are shown for low latent load without space humidity control.

TZ	RZ	TO	WBO	OSA	ZS	ZL	ZR	DTR	BF	SP	LSF	IADP
75	50	95	75	200	22000		4000	20	1	4	2	2

	DB	WB	RH	W	H	DP	ADP
Coil entering	81.3	65.9	44.0	0.01007	30.57	57.4	
Coil leaving	57.5	56.1	91.8	0.00927	23.87	55.1	54.8
Reheat leaving	57.5						
Zone condition	75.0	62.6	50.0	0.00927	28.15	55.1	
Outside air condition	95.0	75.0	39.8	0.01412	38.37	66.7	
Return air	78.0						
Mixer air	80.8	65.7	44.8				

Supply volume, cfm = 1207
Cooling coil load, Btuh = 36,399
Reheat coil load, Btuh = 0

Figure 5. Program input and results are shown for low latent load with space humidity control.

TZ	RZ	TO	WBO	OSA	ZS	ZL	ZR	DTR	BF	SP	LSF	IADP
75	50	95	75	200	22000	11000	4000	20	1	4	2	1

	DB	WB	RH	W	H	DP	ADP
Coil entering	82.4	67.8	47.4	0.01124	32.12	60.4	
Coil leaving	54.1	52.9	92.7	0.00826	21.93	52.0	50.9
Reheat leaving	54.1						
Zone condition	75.0	64.4	56.6	0.01052	29.52	58.6	
Outside air condition	95.0	75.0	39.8	0.01412	38.37	66.7	
Return air	78.6						
Mixed air	81.9	67.6	48.1				

Supply volume, cfm = 1000
Cooling coil load, Btuh = 45,842
Reheat coil load, Btuh = 0

Figure 6. Program input and results are shown for high latent load without space humidity control.

TZ	RZ	TO	WBO	OSA	ZS	ZL	ZR	DTR		BF	SP	LSF	IADP
75.	50.	95.	75.	200.	22000.	11000.	4000.	20.		.1	4.	2	2

	DB	WB	RH	W	H	DP	ADP
Coil entering	82.4	66.4	43.2	0.01024	31.02	57.8	
Coil leaving	49.9	48.7	92.0	0.00700	19.56	47.6	46.3
Reheat leaving	54.1						
Zone condition	75.0	62.6	50.0	0.00927	28.15	55.1	
Outside air condition	95.0	75.0	39.8	0.01412	38.37	66.7	
Return air	78.6						
Mixed air	81.9	66.3	43.9				

Supply volume, cfm = 1000
Cooling coil load, Btuh = 51,593
Reheat coil load, Btuh = 4603

Figure 7. Program input and results are shown for high latent load with space humidity control.

The new version of the Heating and Cooling Load Calculation Program (HCC III) developed by Automated Procedures for Engineering Consultants, Inc. (APEC) has included an enlarged version of this program. As stated at the beginning of this article, many variations of this type of analysis can readily be developed to cover various air conditioning systems. A heating and cooling load calculation program extended to include the psychrometric system analysis will no doubt help engineers to make better and more accurate decisions in air conditioning system design.

References

1. Sun, Tseng-Yao, "Psychrometric Subroutine Uses ASHRAE Algorithms," *Heating/Piping/Air Conditioning,* October 1971.
2. *System Design Manual, Part 1, Load Estimating,* Carrier Corp., Syracuse, N.Y., 1968.

Appendix C

Heat Recovery vs. Economizer Cycle*

Some Insights to Aid You in the Complex Evaluations Required to Determine Whether One or the Other, or a Combination, Will Be Most Cost Effective in a Given Situation

With the great emphasis on energy conservation today, architect-engineer teams in the building construction industry are constantly striving to design buildings and their air conditioning systems to consume the least amounts of energy.

Energy can be conserved by designing well insulated and shaded building envelopes that have minimum heat gains and losses. Nondepleting energy sources can be used to conserve depletable energy sources. Energy can also be conserved by choosing the appropriate energy conserving air conditioning systems. This article discusses various aspects of two commonly used energy conserving schemes in air conditioning system design: refrigeration cycle heat recovery and the economizer cycle. It also investigates the possibility of using a combination of heat recovery and economizer cycle.

*Reprinted with permission from *Heating/Piping/Air Conditioning,* February, 1978.

Characteristics of the Systems

One popular form of heat recovery in building air conditioning is the use of the refrigeration cycle to "pump" rejected heat from spaces requiring cooling to those requiring heat (see Figs. 1 and 2). When rejected heat exceeds the heating load, the excess heat is rejected to the atmosphere through a cooling tower or air cooled condenser. With this scheme, as long as there is rejected heat to satisfy the heating needs, the heat is considered free, since the energy used in the refrigeration cycle is needed for cooling regardless, and no additional energy is used for heating.

An economizer cycle is a design option in various air conditioning schemes that takes advantage of the cooler outside air in mild and cold weather to supplement or satisfy the cooling needs (see Fig. 3). When the outside temperature drops below the cold supply air temperature, the refrigeration cycle can be shut down and the entire cooling load met by using varying amounts of cool outside air. Since the refrigeration plant is shut down, the cooling thus obtained is considered free.

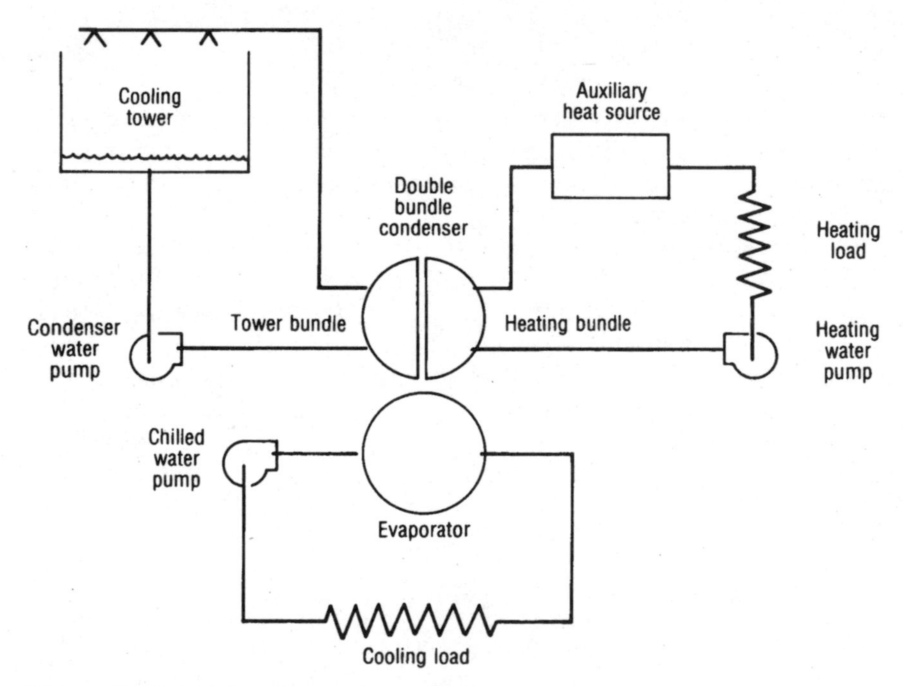

Figure 1. Double bundle condenser heat recovery system.

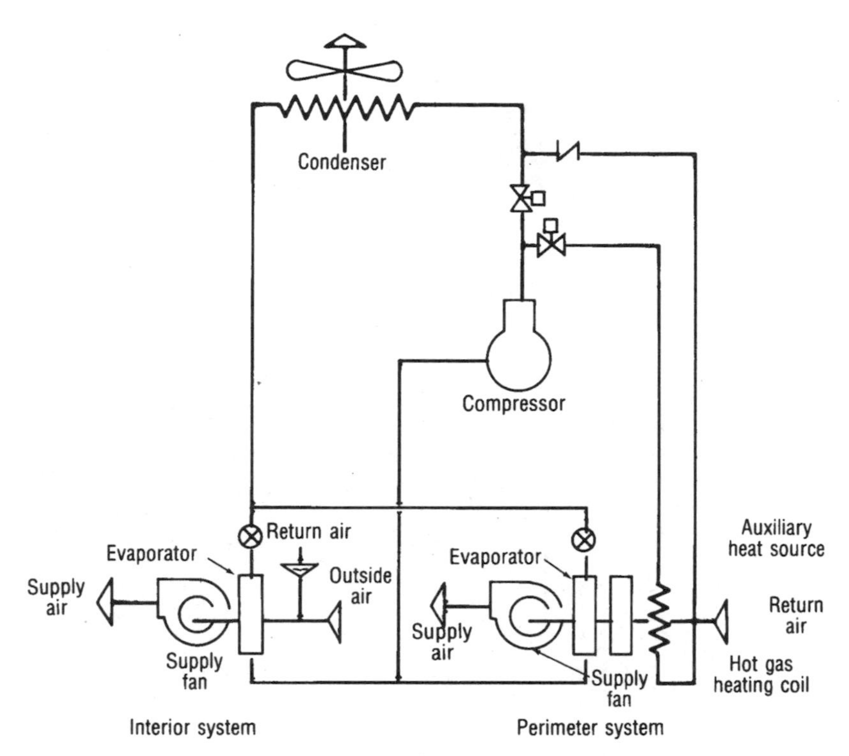

Figure 2. Hot gas heat recovery system.

Both refrigeration cycle heat recovery and the economizer cycle are designed to conserve energy. The concepts of these two energy conservation schemes are basically not compatible, however. The economizer cycle saves cooling energy. It is most effective when the refrigeration plant is shut down. On the other hand, the heat recovery scheme saves heating energy, and it relies on the operation of the refrigeration plant to transfer the rejected heat to satisfy the heating needs.

In reality, however, refrigeration cycle heat recovery and the economizer cycle can be used harmoniously in a building air conditioning system to obtain maximum operating economy. In such an application, the economizer cycle operates in conjunction with the heat recovery plant. Instead of shutting down the refrigeration plant to take full advantage of cool outside air, the economizer cycle is controlled to maintain just enough refrigeration load for the plant to operate such that the rejected heat satisfies the heating requirements.

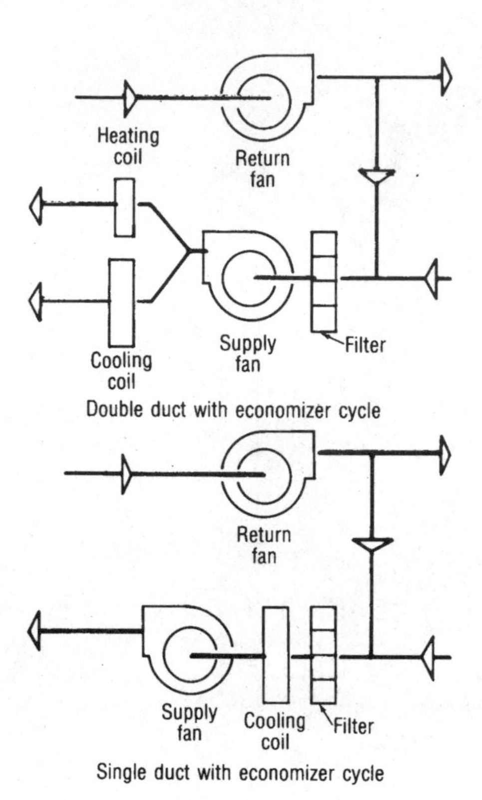

Figure 3. Economizer cycles for single and double-duct systems.

Factors to Consider

Any one of the three schemes, refrigeration cycle heat recovery (which we will subsequently refer to simply as heat recovery), economizer cycle, or combination heat recovery-economizer cycle (which we will refer to as combined system), can be chosen as the air conditioning system for a given building. There are, however, many important and interrelated factors that an engineer must evaluate in the decision making process before he can choose one of these design options. The important factors are:

- Heating and cooling load profiles.
- Availability of energy sources and utility rates.

- Well head energy consumption.
- The justification of the additional initial investments.
- The choice of air handling systems.
- The requirements for humidity control.

Load Profiles, Utility Rates

Generally, if the magnitude of the heating profile of a building is always minimal in comparison with the concurrent cooling profile, such as the load profiles of a well insulated building with high internal loads, the choice may tend to favor the economizer cycle. This is because there is minimum usage of the rejected heat, and much of the cooling energy can be saved with an economizer cycle. On the other hand, if there are significant periods of time when the concurrent heating and cooling profiles have similar magnitudes, the choice may favor the use of heat recovery since the rejected heat can always be used for heating that otherwise would have to be met by some source of new energy. This is especially true of all-electric buildings where resistance heat is used as the heating source.

Depending on the location and the type of building, the availability of energy sources may vary. Under different conditions, coal, oil, natural gas, propane and electricity can all be considered as primary energy sources. Where two energy sources are available, such as gas and electricity, the costs of the sources will influence the choice of systems.

An example using two sets of utility rates and three different levels of heating profiles is set forth below to illustrate the evaluation process concerning the effects of load profiles and utility rates.

Fig. 4 shows a hypothetical cooling profile and three (normal, high, and low) heating profiles, each of which may occur concurrently with the cooling profile. It is assumed that this hypothetical case occurred on a day when the outside air temperature never rose above 50 F, so that the refrigeration plant did not have to operate in the case of the economizer cycle. Further, the following assumptions are made to simplify the illustration:

1. The cooling-only chiller operates at 0.7 KW per ton.
2. The heat recovery chiller operates at 0.88 KW per ton.

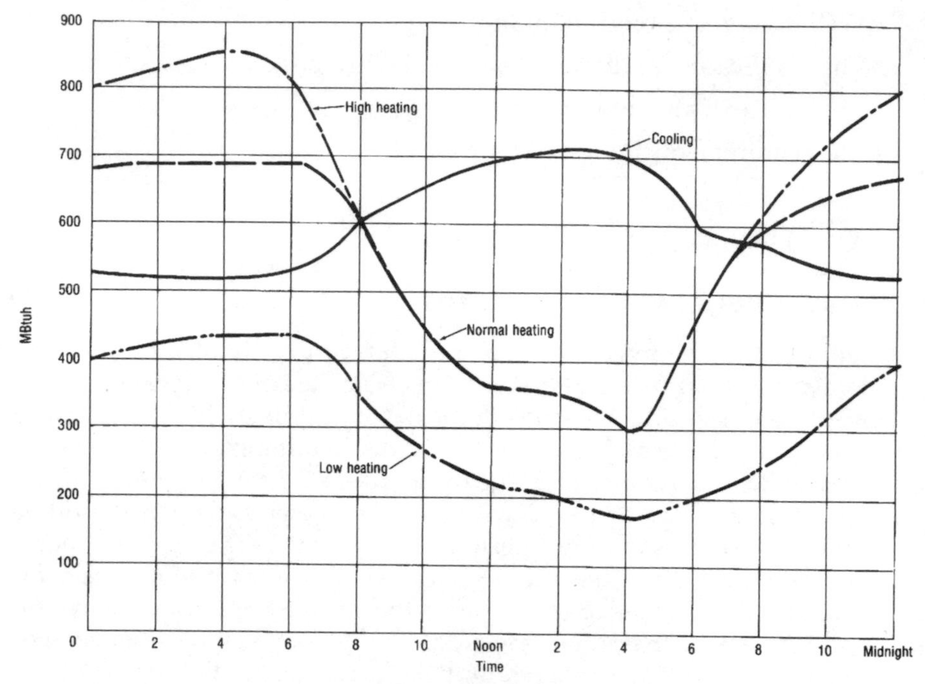

Figure 4. Hypothetical heating and cooling profiles.

3. In the case of an economizer cycle with the refrigeration plant shut down, the power savings of not running the chilled water pump are ignored.
4. Natural gas is used as heating fuel for a hot water boiler in the case of an economizer cycle. Electricity is used as the auxiliary heating source in the cases of heat recovery and combined system.
5. Boiler efficiency is 75 percent; electric heating efficiency is 100 percent.

Table 1 shows, for each hour of the day, the heating and cooling loads, the maximum available rejected heat from the cooling load, the minimum cooling load that will be required to transfer enough rejected heat to match the heating needs of the normal heating profile, the corresponding kilowatts required for the minimum cooling loads, and the kilowatts required for auxiliary heat when there is not enough rejected heat to match the heating needs. Tables 2 and 3 show the same

Table 1 Energy Requirements for Cooling and Normal Heating Profile

Hours	Loads, MBtuh Heating	Loads, MBtuh Cooling	Available rejected heat from cooling load at 15 MBtuh per ton, MBtuh	Minimum cooling required to match heating need, MBtuh	Cooling energy required KW	Auxiliary heating energy KW
1	685	525	656	525	38.5	8.5
2	690	520	650	520	38.1	11.7
3	690	520	650	520	38.1	11.7
4	690	520	650	520	38.1	11.7
5	690	525	656	525	38.5	10.0
6	690	530	663	530	38.9	7.9
7	670	550	688	536	39.3	
8	600	600	750	480	35.2	
9	510	630	788	408	29.9	
10	450	650	813	360	26.4	
11	390	670	838	312	22.9	
Noon	360	690	863	288	21.1	
1	365	705	881	292	21.4	
2	350	710	888	280	20.5	
3	330	710	888	264	19.4	
4	300	700	875	240	17.6	
5	360	670	838	288	21.1	
6	460	600	750	368	27.0	
7	550	580	725	440	32.3	
8	600	570	713	480	35.2	
9	630	550	688	504	37.0	
10	650	540	675	520	38.1	
11	660	545	681	528	38.7	
Midnight	670	530	662	530	38.9	2.3
Totals	13,040 MBtu	14,340 MBtu			752.2 KWH	63.8 KWH

for the high heating profile and for the low heating profile, respectively.

Two sets of utility rates are used for the comparison, as shown in Table 4. These rates are taken from 1977 records for two relatively large facilities located in cities 50 miles apart. Rate A tends to favor natural gas because of the relatively higher cost of electricity, whereas Rate B favors electricity because of its unusually low cost relative to the gas cost. Despite the fact that these rates have been obsoleted by today's mounting escalation, they nevertheless give a good relative

Table 2 Energy Requirements for Cooling and High Heating Profile

Hours	Loads, MBtuh Heating	Loads, MBtuh Cooling	Available rejected heat from cooling load at 15 MBtuh per ton, MBtuh	Minimum cooling required to match heating need, MBtuh	Cooling energy required KW	Auxiliary heating energy KW
1	815	525	656	525	38.5	46.6
2	830	520	650	520	38.1	52.8
3	840	520	650	520	38.1	55.7
4	850	520	650	520	38.1	58.6
5	840	525	656	525	38.5	54.0
6	800	530	663	530	38.9	40.2
7	710	550	688	550	40.3	6.5
8	600	600	750	480	35.2	
9	510	630	788	408	29.9	
10	450	650	813	360	26.4	
11	390	670	838	312	22.9	
Noon	360	690	863	288	21.1	
1	365	705	881	292	21.4	
2	350	710	888	280	20.5	
3	330	710	888	264	19.4	
4	300	700	875	240	17.6	
5	360	670	838	288	21.1	
6	460	600	750	368	27.0	
7	550	580	725	440	32.3	
8	620	570	713	480	35.2	
9	690	550	688	550	40.3	0.6
10	730	540	675	540	39.6	16.1
11	770	545	681	545	40.0	26.1
Midnight	800	530	662	530	38.9	40.5
Totals	14,320 MBtu	14,340 MBtu			759.3 KWH	397.7 KWH

comparison of the effect of utility costs. For convenience, the electricity rates are given in dollars per kilowatt-hours, which includes the demand charges and fuel adjustment factors.

Energy costs were calculated for each design option using different sets of utility rates for the three sets of heating and cooling profiles. Table 5 shows the calculation procedures used for Rate A and the normal heating profile; calculations for other combinations of rates and profiles followed the same procedures. The summary of the energy cost comparison for our hypothetical day is presented in Table 6.

Table 3 Energy Requirements for Cooling and Low Heating Profile

Hours	Loads, MBtuh Heating	Loads, MBtuh Cooling	Available rejected heat from cooling load at 15 MBtuh per ton, MBtuh	Minimum cooling required to match heating need, MBtuh	Cooling energy required KW	Auxiliary heating energy KW
1	415	525	656	332	24.3	
2	420	520	650	336	24.6	
3	425	520	650	340	24.9	
4	430	520	650	334	24.5	
5	435	525	656	348	25.5	
6	440	530	663	352	25.8	
7	410	550	688	328	24.1	
8	350	600	750	280	20.5	
9	300	630	788	240	17.6	
10	270	650	813	216	15.8	
11	240	670	838	192	14.1	
Noon	220	690	863	176	12.9	
1	210	705	881	168	12.3	
2	200	710	888	160	11.7	
3	180	710	880	144	10.6	
4	170	700	875	136	10.0	
5	180	670	838	144	10.6	
6	200	600	750	160	11.7	
7	225	580	725	180	13.2	
8	250	570	713	200	14.7	
9	280	550	688	224	16.4	
10	330	540	675	264	19.4	
11	370	545	681	296	21.7	
Midnight	400	530	662	320	23.5	
Totals	7,350 MBtu	14,340 MBtu			430.4 KWH	

Table 4 Utility Rates Used in Example

	Rate A	Rate B
Natural gas	$1.06 per Mcf	$1.14 per Mcf
Electricity	$0.033 per KWH	$0.015 per KWH

Table 5 Energy Cost Calculation for Rate A and Normal Heating Profile (Calculation Procedures Were Similar for Other Rate and Profile Combinations)

CONVENTIONAL SYSTEM

Heating: $\dfrac{13{,}040 \text{ MBtu}}{0.75 \text{ Eff.}} \times \dfrac{\$1.06/\text{Mcf}}{1000 \text{ Btu/cf}} = \$18.43$

Cooling: $\dfrac{14{,}340 \text{ MBtu}}{12 \text{ MBtu/ton-hr}} \times 0.7 \text{ KW/ton} \times \$0.033/\text{KWH} = \$27.60$

Total: $18.43 + $27.60 = $46.03

ECONOMIZER CYCLE

Total: $18.43 + 0 = $18.43 (heating as above; no cooling)

HEAT RECOVERY

Cooling: $\dfrac{14{,}340 \text{ Btu}}{12 \text{ MBtu/ton-hr}} \times 0.88 \text{ KW/ton} \times \$0.033/\text{KWH} = \$34.70$

Heating: 63.8 KWH × $0.033/KWH = $2.11 (auxiliary)

Total: $34.70 + $2.11 = $36.81

COMBINED SYSTEM

Cooling: 752.2 KWH × $0.033/KWH = $24.82

Heating: $2.11 (auxiliary)

Total: $24.82 + $2.11 = $26.93

Table 6 reveals some interesting results that demonstrate the complexities of such evaluations. They show that with Rate A, the energy costs for an economizer cycle are always lower than those for heat recovery or for a combined system. This indicates that with the higher cost of electricity, it is more expensive to "pump" the rejected heat to satisfy the heating needs than to use the comparatively lower cost natural gas to do the heating directly. In fact, with the high heating profile, the heat recovery with electric resistance heating offers practically no energy cost savings in comparison with the conventional system.

The combined system trims the energy usage of the heat recovery system. The energy costs in this case, although significantly reduced, are still higher than those for an economizer cycle.

Table 6 Summary of Heating and Cooling Energy Costs for Conventional and Energy Saving Systems for Various Combinations of Utility Rates (Table 4) and Load Profiles (Fig. 4), Calculated as Shown in Table 5

	Rate A Heating profile			Rate B Heating profile		
System	High	Normal	Low	High	Normal	Low
Conventional	$47.80	$46.00	$38.00	$34.30	$32.40	$23.70
Economizer	20.20	18.40	10.40	21.60	19.80	11.20
Heat recovery	47.80	36.80	34.70	21.70	16.70	15.80
Combination	38.20	26.90	14.20	17.40	12.20	6.50

With Rate B, the energy costs for using heat recovery become less than those for an economizer cycle as long as the heating and cooling profiles have similar magnitudes. This indicates that with cheaper electricity it becomes economical to pump the rejected heat to satisfy the heating load.

When the heating profile is comparatively higher than the cooling profile, so that more auxiliary heat must be used to satisfy the heating load, the energy cost difference between the economizer cycle and heat recovery becomes minimal. This is primarily because the electric heating used as the auxiliary heating source with heat recovery is not as cost effective as the gas heating used with the economizer cycle.

When the heating profile is comparatively lower than the concurrent cooling profile—in other words, when the rejected heat becomes much greater than the actual heating needs—the economizer cycle again becomes more economical than the heat recovery system.

In the case of Rate B, the trimming action of the combined system makes it the most energy efficient choice among the three options for all profiles.

Well Head Energy Consumption

Utility costs include not only the well head costs for various forms of usable energy; they also reflect the processing costs involved in making energy sources into usable form. One should recognize, therefore,

Table 7 Summary of 24 hr Energy Consumption Data for Conventional and Energy Saving Systems for Various Combinations of Assumed Power Generation Conversion Factors and Load Profiles (from Fig. 4)

	Energy Consumption, MBtu					
	@ 11,600 Btu per KWH[1]			@ 10,239 Btu per KWH[2]		
	Heating profile			Heating profile		
System	High	Normal	Low	High	Normal	Low
Conventional	28,800	27,090	19,500	27,660	25,960	18,370
Economizer	19,090	17,390	9,800	19,090	17,390	9,800
Heat recovery	16,810	12,940	12,200	14,840	11,420	10,770
Combination	13,440	9,470	4,990	11,860	8,360	4,410

[1]GSA value for power generation conversion factor.
[2]State of California value for power generation conversion factor.

that the lowest energy *cost* option may not be the lowest energy *consumption* option.

The General Services Administration uses 11,600 Btu per KWH as a conversion factor for electrical energy generation. The State of California uses 10,239 Btu per KWH as a conversion factor in its code. Table 7 shows the energy consumptions of the three system operations under various load conditions based on each of these factors. The table shows that in almost all cases the heat recovery and combined systems use the least amounts of energy.

The Heating Profiles

In the above example, it was shown that the heating profiles play an important role in determining which of the three options should be chosen as the most energy effective. For this discussion, it is important to define what constitutes a heating profile and how it differs from the conventional heating load calculation.

The heating profiles used in such an evaluation must reflect the net heating needs; *i.e.*, the actual heating loads imposed on the heating apparatus. For example, in a conventional heating load calculation, we never take credit for the solar radiation that may neutralize part of the heat losses during the daylight hours. We also ignore the heat of lights and the heat from human bodies, which in reality help to heat a building during the hours the building is in operation.

All of these factors—heat from the sun, lights, and people—tend to reduce the magnitude of the heating profile. Unless they are properly accounted for in constructing a profile, the results of an evaluation will be distorted and lose validity.

Since both heating and cooling profiles are functions of varying weather conditions, solar radiation, and internal loads, it is sometimes difficult and time consuming to evaluate these intricate problems manually. Engineers often resort to computer programs to perform these complex evaluations.

Thermal Storage

The above example also illustrates that heat recovery and the combined system work best when the cooling and heating profiles have similar magnitudes. The use of auxiliary heat should be minimized, especially when electricity is used as the source.

The heavy heating load in a building generally occurs in the early morning hours or during the morning warmup periods. Unfortunately, the cooling load during these periods is generally minimal if existing at all. On the other hand, the maximum cooling load generally occurs in mid-afternoon when the heating load is at its minimum. To alleviate these mismatching conditions and to prevent the extensive use of auxiliary heat, many heat recovery systems are equipped with thermal storage tanks.

Initial Investment

It cannot be denied that many existing air conditioning systems waste varying amounts of energy. A major reason is that in the past the primary goal in building air conditioning system design was to minimize the installed cost. A vivid illustration of this can be found in the previous example. The conventional system, without any additional equipment for some form of energy conservation, undoubtedly is the lowest cost system from an initial investment viewpoint. The energy cost of the conventional system under the given weather conditions, however, is more than double that of the most energy conserving system.*

*It should be noted that the drastic energy cost differences shown here result from the fact that the weather conditions used in the example favor the energy conserving designs. Cost differences would not be as pronounced during mild or warm weather conditions.

Engineers are often asked to justify the additional expenditures required for energy conserving systems. In the case of an economizer cycle, these expenditures include additional return air fans, dampers, controls, and possibly additional space in the mechanical rooms. The additional expenditures for a heat recovery system include condenser modification, auxiliary heating, storage tanks if required, and larger piping and deeper coils in the heating system due to the temperature limitations in the heat recovery system. A combined system involves expenditures for the features of both of the above systems as well as for additional controls.

Economizer cycles and heat recovery are now required or recommended by many local code authorities as an energy conservation measure. Where this is the case, the justification of initial investments for such systems becomes academic. The California State Energy Code, for example, *encourages* the use of heat recovery, but it *requires* that an economizer cycle be incorporated into the air conditioning system design if the air handling system capacity exceeds 5000 cfm.*

It can be seen in Table 6 that the majority of the energy cost savings over the conventional design is achieved by the use of either an economizer cycle or heat recovery. The additional energy cost savings due to the use of a combined system are relatively minor. Since the initial investment for a combined system is much higher than for either of the individual systems, it is sometimes difficult to justify its use, unless the criteria in system selection heavily favor energy conservation with less concern about initial investment.

Air Handling System Choice

The selection of air handling systems must be integrated into the evaluation to choose among various energy conserving schemes. While the use of heat recovery is practically independent of air handling system selection, the use of an economizer cycle has a definite effect on air handling system selection.

An economizer cycle used in conjunction with a double-duct or mul-

*It is interesting to note that in California one must use an economizer cycle when the size of the system exceeds the code limit even if the building is located in an area where the utility rates favor the use of heat recovery. If the engineer wants to use heat recovery instead of an economizer cycle because of the characteristics of the local utility rates or for other compelling engineering reasons, he or she must submit proof to the code authorities that an economizer cycle is not as economical as heat recovery.

tizone system will not perform as effectively as an economizer cycle coupled with a single-duct reheat or single-duct variable volume system. When an economizer cycle is used with a conventional double-duct or multizone system, the heating profile becomes a function of air handling system operation. When the outside air temperature drops below the cold air supply temperature, the economizer cycle will always try to maintain the mixed air temperature (outside air and return air) to match the cold air supply temperature by introducing more outside air than otherwise required. Heating is penalized because the heating coil must then heat the air from the cold air supply temperature instead of from the warmer mixed air temperature it would work with if the system did not include an economizer cycle. For this reason, a heat recovery system is often a better energy conservation scheme for double-duct systems.

To avoid this heating penalty, some double-duct systems are designed with separate fans for hot and cold air supply. The economizer cycle is incorporated into the cold air system only. The hot air system uses 100 percent return air or a fixed minimum amount of outside air.

The use of an economizer cycle does not penalize a single-duct system. With a single-duct reheat system, the reheat coil entering air temperature is the same whether or not the system uses an economizer cycle. With a dual conduit system or VAV system with perimeter radiation, the heating system is completely independent of the cooling system, and an economizer cycle in the cooling system therefore has no effect on the heating system.

Humidity Requirements

The humidification load in an air conditioning system is primarily a function of the amount of outside air required. Since an economizer cycle relies on an increased amount of outside air to conserve cooling energy, the humidification load will increase when the increased amount of cool outside air is drier than the humidity level to be maintained in the space.

Where an economizer cycle is used for conditioned spaces requiring high humidity levels, such as computer rooms and hospital surgical suites, it is possible that for a given set of utility rates, the cost of additional humidification may exceed the cooling cost saving. In this case, a heat recovery system naturally becomes the preferred energy conserving scheme.

Conclusion

There are no simple rules for deciding what energy conserving scheme should be used in a given building. In addition to the three approaches considered here, many other design concepts can be used to conserve energy effectively.

As illustrated in this article, the evaluation process can be so complex that computer programs may be needed to help speed up what otherwise would be an almost impossible task. In such cases, the engineer should recognize that because of the complexity involved, he must have a complete understanding of, and agreement with, the algorithms used in the computer programs to be able to claim that he, not the author of the program, performed the evaluation. It is mainly for this reason that Automated Procedures for Engineering Consultants, Inc. (APEC), has prepared its energy analysis program, Energy Simulation Program—Level I. Released by APEC in 1978,* ESP-I is a program written by engineers and for engineers.

Energy conservation forces the building industry and design engineers to think beyond the initial installation of air conditioning systems. Air conditioning systems designed for buildings of the future will no longer be based on initial investment considerations alone. Energy conserving options are definitely becoming an integral part of the decision-making process of selecting building air conditioning systems.

*See "Design Energy Efficient Buildings with ESP," by Charles Kalasinsky and Francis Ferreira, *Heating/Piping/Air Conditioning,* October 1978, p. 75.

Appendix D

Design Experience with Indirect Evaporative Cooling*

A thorough engineering analysis of energy conservation options is generally always required. What looks good on a preliminary basis may not hold up under scrutiny.

This article discusses a design experience of evaluating the possibility of using indirect evaporative cooling in a 135,000-ft^2 seven-story research laboratory building on a university campus. The use of indirect evaporative cooling can possibly help precool the large quantity of required outside air to conserve refrigeration energy and reduce the installed refrigeration capacity for the building.

Evaporative cooling is an adiabatic process that cools air by evaporating water into an air stream. The latent heat of evaporation lowers the dry-bulb temperature of the air to produce the cooling sensation. No heat is removed from the airstream in an evaporative cooling process.

Indirect evaporative cooling, on the other hand, is a cooling process utilizing the dry-bulb depression of evaporative cooling to cool an airstream through an air-to-air heat exchanger. Indirect evaporative cooling is a sensible cooling process; i.e., sensible heat is removed from the airstream. When used in conjunction with mechanical refrig-

*Reprinted with permission from *Heating/Piping/Air Conditioning*, January 1988.

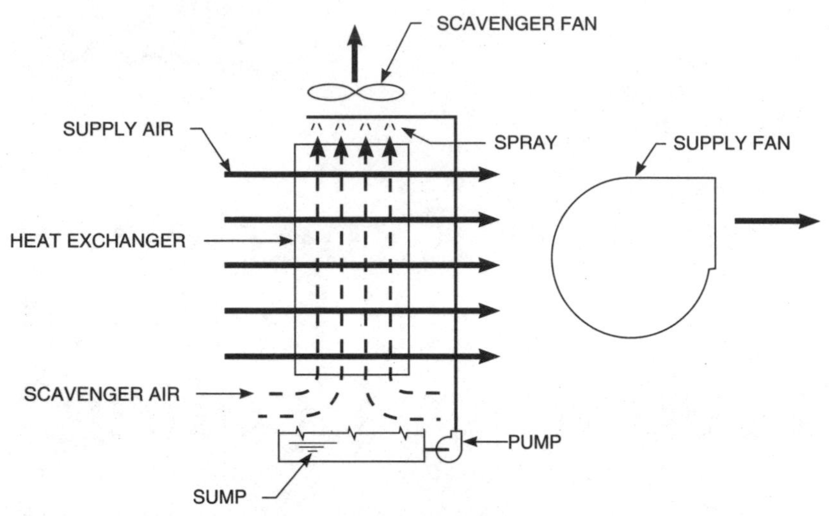

Figure 1. Indirect evaporative cooling unit.

eration cooling, the indirect evaporative cooling process can save refrigeration energy. Figure 1 shows a schematic diagram of an indirect evaporative cooling system. Figure 2 shows an associated psychrometric plot.

The typical commercially available indirect evaporative cooling system is coupled with a direct evaporative cooling process to maximize the depression of the dry-bulb temperature. The system is commonly referred to as an *indirect-direct evaporative cooling system.* In areas where the weather is hot and dry, this type of air-conditioning system is very popular and effective. For installations where close temperature control is not essential, this system often replaces mechanical refrigeration. This article is limited to the discussion of the indirect evaporative cooling application only.

The conclusion of this evaluation is that indirect evaporative cooling cannot be justified for this specific project. However, the device can be a very effective energy saving tool under certain circumstances.

The Environment

The campus is located in a coastal region of the Los Angeles area in a mild climatic zone. The ASHRAE 0.1 percent design dry-bulb and mean coincidental wet-bulb temperatures are 91°F and 67°F, respec-

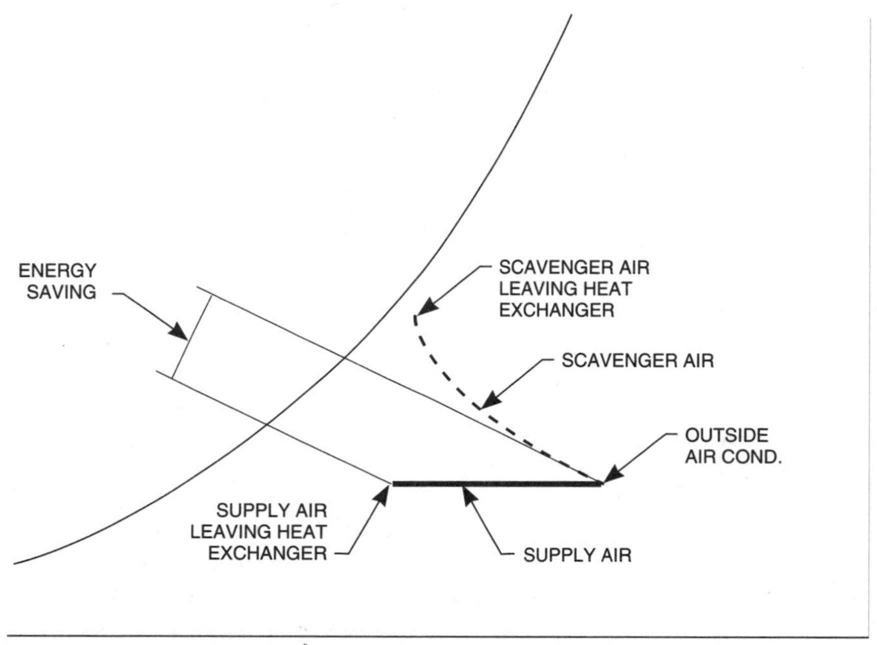

Figure 2. Psychrometric plot of indirect evaporative cooling process.

tively. The laboratory systems will be operating 24 hours per day. Because the laboratory will require close temperature control at all times, mechanical refrigeration will be required. The building will be served from a large central cooling plant designed to serve the majority of the buildings on campus through a distribution loop.

The central plant uses centrifugal chillers with low kilowatt per ton power requirements. The overall plant coefficient of performance (*COP*), including the chillers, chilled-water pumps, condenser water pumps, and cooling tower fans, is 3.9, which translates to 0.9 KW per ton. The average electricity cost for the campus is approximately 7¢ per KWH. The water cost is $1.10 per 1000 gallons.

A False Start

In this age of "energy conservation," engineers get very excited when an idea comes to mind that promises a large energy saving. This experience was no exception. The project requires approximately 150,000 cfm of outside air. Under the outside design condition, 580 tons of refrigeration will be required to cool this outside air to the supply air condition. If this large amount of outside air can be "free-cooled" to a lower temperature before entering the chilled-water cooling coil, the process will conserve refrigeration energy, and the cooling plant size can be reduced.

An indirect evaporative cooler with relatively high thermal efficiency, offered by Manufacturer A, was selected for this study. The cooler can cool the outside air from the 91/67°F design outside condition to 71.8/60.5°F. Under this condition, only 326 tons of refrigeration are needed to cool the air down to the supply air temperature, a reduction of 254 tons.

The estimated installation cost of the precooler assembly is $200,000. An estimated saving for the refrigeration capacity avoidance is $350 per ton. The precooler cost can be justified if the combined savings of a smaller chiller plant, reduced investment, and operating costs can recover the incremental cost difference—precooler minus credit for smaller chiller plant—in a reasonable time.

A small computer program was quickly written to calculate the potential energy saving. Monthly bin weather data from *Engineering Weather Data* by the Department of the Armed Forces were used for the calculation. Equations were developed to calculate the dry-bulb and wet-bulb temperatures leaving the precooler. For a given KW per ton input, energy consumption in KWH can be calculated for a system with and without a precooler. Percent of energy saving in each temperature bin using the precooler was also calculated to give the engineer a feel of how the saving related to the weather variation.

A partial output of the program for four months is shown in Table 1. Knowing the cost of electricity, the engineer can calculate the annual refrigeration cost savings using the precooler. With the cost of the precooler and the estimated cost savings of the reduced cooling plant tonnage, one can readily calculate a simple payback period. Table 2 shows this result. The short payback period of 3.85 yr was reasonable. Serious consideration was given to arranging the space requirement to install a precooling apparatus.

The Complete Picture

In a closer look at the system, however, several important factors must also be taken into consideration: The insertion of the precoolers in the outside air stream adds to system static pressure. The fans and pumps for the evaporative cooler use energy to operate, which dimin-

Table 1. Partial Output of Potential Energy Saving Program for a 4-Month Period

OSA TEMP.		NO. OF HOURS	PRECOOL LVG.		COOLING ENERGY KWH		PERCENT SAVING W/ PRECOOLING
DB	WB		DB	WB	CONVENTIONAL	W/ PRECOOL	
SEPT							
102.	72.0	1	78.0	64.6	622.	325.	47.7
97.	69.0	2	74.6	61.7	1040.	485.	53.3
92.	69.0	1	73.6	63.0	496.	268.	45.9
87.	68.0	6	71.8	63.0	2738.	1608.	41.2
82.	68.0	21	70.8	64.4	9633.	6715.	30.3
77.	66.0	63	68.2	63.1	23935.	17059.	28.7
72.	64.0	144	65.6	61.8	43846.	32422.	26.1
67.	62.0	230	63.0	60.6	53415.	42015.	21.3
62.	60.0	221	60.4	59.4	36019.	31638.	12.2
57.	56.0	30	56.2	55.7	929.	633.	31.9
OCT.							
102.	67.0	1	74.0	57.2	582.	235.	59.6
97.	65.0	1	71.4	55.7	520.	203.	61.0
92.	63.0	3	68.8	54.3	1374.	512.	62.7
87.	62.0	4	67.0	54.4	1584.	594.	62.5
82.	62.0	13	66.0	56.0	4344.	1770.	59.3
77.	63.0	26	65.8	59.0	7079.	3475.	50.9
72.	62.0	84	64.0	59.1	19332.	11023.	43.0
67.	61.0	198	62.2	59.2	38854.	27091.	30.3
62.	58.0	264	58.8	56.8	24994.	14551.	41.8
57.	55.0	127	55.4	54.3	3143.	629.	80.0
52.	50.0	22	50.4	49.3	0.	0.	.0
NOV.							
92.	60.0	1	66.4	49.7	458.	141.	69.2
87.	59.0	4	64.6	49.9	1584.	475.	70.0
82.	56.0	10	61.2	47.1	3341.	767.	77.0
77.	56.0	20	60.2	48.9	5445.	1287.	76.4
72.	57.0	35	60.0	52.1	7363.	2166.	70.6
67.	56.0	88	58.2	52.4	13068.	3485.	73.3
62.	55.0	192	56.4	52.7	16632.	3326.	80.0
57.	52.0	215	53.0	50.3	5321.	0.	100.0
52.	48.0	115	48.8	46.5	0.	0.	.0
47.	43.0	37	43.8	41.4	0.	0.	.0
42.	38.0	4	38.8	36.2	0.	0.	.0
DEC.							
82.	56.0	5	61.2	47.1	1671.	384.	77.0
77.	54.0	10	58.6	45.8	2723.	446.	83.6
72.	53.0	21	56.8	46.2	4418.	468.	89.4
67.	53.0	48	55.8	48.1	7128.	475.	93.3
62.	53.0	128	54.8	49.9	11088.	0.	100.0
57.	52.0	226	53.0	50.3	5594.	0.	100.0
52.	48.0	190	48.8	46.5	0.	0.	.0
47.	43.0	91	43.8	41.4	0.	0.	.0
42.	38.0	21	38.8	36.2	0.	0.	.0
37.	32.0	2	33.0	29.8	0.	0.	.0
TOTAL		8753			1051313.	639286.	39.2

ishes the thermal energy savings on the indirect precooler side. The evaporation process of the cooler also uses a significant amount of water. These influencing factors must be included in the overall evaluation.

The computer program was modified to include these additional factors. The added horsepower of the supply fan was calculated using the static pressure across the precooler at the rated air quantity. The energy consumption for this added horsepower is accumulated. This energy usage is forever present as long as the air-conditioning system is in operation. The energy consumption for the fans and pumps of the evaporative cooler, which is a function of the operating time of the cooler, is also accumulated. The water consumption, which is a function of the weather and the cooler efficiency, is also calculated.

Table 2. Simple Payback Period Can Be Calculated with the Cost of the Precooler and the Estimated Cost Saving of the Reduced Cooling Plant Tonnage.

```
ENERGY CONSUMPTION FOR CONVENTIONAL SYSTEM      =    1051313. KW-HR.
ENERGY CONSUMPTION WITH PRECOOLING SYSTEM       =     639286. KW-HR.
ENERGY CONSUMPTION SAVING FOR PRECOOLING        =     412028. KW-HR.

ANNUAL ENERGY COST SAVING FOR PRECOOLING        =      28842. DOLLARS

COST OF THE PRECOOLER                           =     200000. DOLLARS
COST SAVING OF 254.0 TON AT $ 350.0/TON         =      88900. DOLLARS

NET INVESTMENT COST                             =     111100. DOLLARS

SIMPLE PAYBACK PERIOD                           =       3.85 YEARS

       ******************** INPUT DATA ********************

       SITE: LOS ANGLES AIRPORT

       OUTSIDE AIR QUANTITY                 =  150000 CFM
       ROOM DESIGN DRY BULB                 =   75.00 DEGREE
       ROOM DESIGN WET BULB                 =   62.50 DEGREE
       COOLING COIL LEAVING DRY BULB        =   55.00 DEGREE
       COOLING COIL LEAVING WET BULB        =   54.50 DEGREE
       KW PER REFRIGERATION TON             =     .90 KW/TON
       ELECTRICITY COST                     =     .07 $/KWH

       ****************************************************


       COOLING IS CONSIDERED SENSIBLE ONLY WHEN HUMIDITY RATIO
       OF THE OUTSIDE AIR IS LOWER THAN THAT OF THE ROOM DESIGN.
       NO HUMIDIFICATION IS CONTEMPLATED.
```

A partial output of the modified program for the same four months is shown in Table 3. The summary of the results is shown in Table 4. With the added factors, the economic picture changed completely. The simple payback period increased from 3.85 to 13.98 yr. A 14-yr payback period is much too long to justify the use of the precooler.

Note that in this particular case the cost of water does not become a part of the evaluation. The cooling tower of the central plant uses the latent heat of water evaporation to remove the rejected heat much the same as the water usage of the evaporative cooler. This calculation also did not include any added intangible maintenance costs associated with the use of the precooler. Naturally, with any maintenance cost consideration, the payback period will be even longer.

The indirect evaporative cooling system is most effective in hot and dry climates where a large differential between dry- and wet-bulb temperature prevails. The system is also more attractive where the existing mechanical refrigeration system has a lower overall *COP*, i.e., higher KW per ton.

Table 3. Energy Saving Program Shown in Table 1, Modified to Include Additional Influencing Factors (See text)

OSA TEMP. DB	OSA TEMP. WB	NO. OF HRS.	COOLING ENERGY KWH NORMAL	COOLING ENERGY KWH PRE-COOL	ANCILLARY KWH EXCHG'R	ANCILLARY KWH SCAVG'R	PRE-COOL TOTAL KWH	WATER CONSUMED GALLONS
SEPT								
102.	72.0	1	622.	325.	15.	24.	364.	567.
97.	69.0	2	1040.	485.	30.	47.	562.	1055.
92.	69.0	1	496.	268.	15.	24.	307.	434.
87.	68.0	6	2738.	1608.	90.	141.	1840.	2152.
82.	68.0	21	9633.	6715.	314.	495.	7524.	5561.
77.	66.0	63	23935.	17059.	942.	1485.	19486.	13088.
72.	64.0	144	43846.	32422.	2153.	3395.	37970.	21724.
67.	62.0	230	53415.	42015.	3439.	5422.	50875.	21656.
62.	60.0	221	36019.	31638.	3304.	5210.	40152.	8313.
57.	56.0	30	929.	633.	449.	707.	1768.	562.
OCT.								
102.	67.0	1	582.	235.	15.	24.	274.	655.
97.	65.0	1	520.	203.	15.	24.	241.	598.
92.	63.0	3	1374.	512.	45.	71.	628.	1624.
87.	62.0	4	1584.	594.	60.	94.	748.	1868.
82.	62.0	13	4344.	1770.	194.	306.	2270.	4866.
77.	63.0	26	7079.	3475.	389.	613.	4477.	6838.
72.	62.0	84	19332.	11023.	1256.	1980.	14259.	15786.
67.	61.0	198	38854.	27091.	2960.	4668.	34718.	22334.
62.	58.0	264	24994.	14551.	3947.	6223.	24722.	19795.
57.	55.0	127	3143.	629.	1899.	2994.	5521.	4748.
52.	50.0	22	0.	0.	329.	0.	329.	0.
NOV.								
92.	60.0	1	458.	141.	15.	24.	180.	594.
87.	59.0	4	1584.	475.	60.	94.	629.	2081.
82.	56.0	10	3341.	767.	150.	236.	1152.	4818.
77.	56.0	20	5445.	1287.	299.	471.	2057.	7799.
72.	57.0	35	7363.	2166.	523.	825.	3514.	9785.
67.	56.0	88	13068.	3485.	1316.	2074.	6875.	18050.
62.	55.0	192	16632.	3326.	2870.	4526.	10723.	25074.
57.	52.0	215	5321.	0.	3214.	5068.	8283.	20065.
52.	48.0	115	0.	0.	1719.	0.	1719.	0.
47.	43.0	37	0.	0.	553.	0.	553.	0.
42.	38.0	4	0.	0.	60.	0.	60.	0.
DEC.								
82.	56.0	5	1671.	384.	75.	118.	576.	2409.
77.	54.0	10	2723.	446.	150.	236.	831.	4258.
72.	53.0	21	4418.	468.	314.	495.	1277.	7390.
67.	53.0	48	7128.	475.	718.	1132.	2324.	12472.
62.	53.0	128	11088.	0.	1914.	3017.	4931.	21426.
57.	52.0	226	5594.	0.	3379.	5328.	8706.	21028.
52.	48.0	190	0.	0.	2041.	0.	2841.	0.
47.	43.0	91	0.	0.	1360.	0.	1360.	0.
42.	38.0	21	0.	0.	314.	0.	314.	0.
37.	32.0	2	0.	0.	30.	0.	30.	0.
TOTAL		8753	1051313.	639286.	130862.	167655.	937803.	847741.

In this particular instance, the 7¢ per KWH electricity cost for the campus is lower than that for most of the commercial buildings in the same area. This is largely due to the bulk usage and relatively stable load on the campus. Since the reason for using an indirect evaporative cooler is to save energy cost, the lower electricity cost has the tendency to lengthen the payback period.

It becomes obvious that the mild weather conditions at the project site, the higher plant efficiency of the mechanical refrigeration system, and the favorable electricity cost to the campus are the basic contributing factors that cause the long payback period.

Table 4. Summary of Results from Program Listed in Table 3

```
ENERGY CONSUMPTION FOR CONVENTIONAL SYSTEM     =   1051313. KW-HR.
ENERGY CONSUMPTION WITH PRECOOLING SYSTEM      =    937803. KW-HR.
ENERGY CONSUMPTION SAVING FOR PRECOOLING       =    113510. KW-HR.

ANNUAL ENERGY COST SAVING FOR PRECOOLING       =      7946. DOLLARS

COST OF THE PRECOOLER                          =    200000. DOLLARS
COST SAVING OF 254.0 TON AT $ 350.0/TON        =     88900. DOLLARS

NET INVESTMENT COST                            =    111100. DOLLARS

SIMPLE PAYBACK PERIOD                          =      13.98 YEARS

     ******************** INPUT DATA ********************

     SITE:  LOS ANGLES AIRPORT

     OUTSIDE AIR QUANTITY                =  150000 CFM
     ROOM DESIGN DRY BULB                =   75.00 DEGREE
     ROOM DESIGN WET BULB                =   62.50 DEGREE
     COOLING COIL LEAVING DRY BULB       =   55.00 DEGREE
     COOLING COIL LEAVING WET BULB       =   54.50 DEGREE
     KW PER REFRIGERATION TON            =     .90 KW/TON
     ELECTRICITY COST                    =     .07 $/KWH
     AIR-TO-AIR HEAT EXCHANGER SP        =     .68 INCH
     FAN EFFICIENCY                      =   80.00%
     SCAVENGER FANS AND PUMP             =   31.60 HP
     SCAVENGER SHUTOFF DRY BULB (OSA)    =   55.00 DEGREE

     ****************************************************

     COOLING IS CONSIDERED SENSIBLE ONLY WHEN HUMIDITY RATIO
     OF THE OUTSIDE AIR IS LOWER THAN THAT OF THE ROOM DESIGN.
     NO HUMIDIFICATION IS CONTEMPLATED.
```

Additional Considerations

Many of these indirect coolers are applied to reduce the size of or to replace the rooftop air cooled packaged air conditioning units. The *COP*s of these rooftop units are generally much lower than that of a central plant utilizing centrifugal chillers. The KW per ton figure for the rooftop units is generally in the range of 1.3 to 1.5. The computer program written for this project can be readily used to analyze the effect of the higher energy requirements. Table 5 shows a calculation with 1.4 KW per ton energy input. The payback period drops to 5.37 yr, a considerable reduction from 13.98 yr for the higher efficiency plant.

Note that with the air-cooled rooftop units, the cost of water becomes a part of the cost analysis, since the air-cooled system does not involve water usage. Furthermore, since the air cooled condensing units associated with the rooftop units are less expensive than the cen-

Table 5. Program Written to Accommodate Higher Energy Requirement of 1.4 kW/ton Energy Input

```
ENERGY CONSUMPTION FOR CONVENTIONAL SYSTEM      =   1635376. KW-HR.
ENERGY CONSUMPTION WITH PRECOOLING SYSTEM       =   1292961. KW-HR.
ENERGY CONSUMPTION SAVING FOR PRECOOLING        =    342415. KW-HR.
WATER CONSUMPTION FOR PRECOOLER                 =    847741. GALLONS

ANNUAL ENERGY COST SAVING FOR PRECOOLING        =     23037. DOLLARS

COST OF THE PRECOOLER                           =    200000. DOLLARS
COST SAVING OF 254.0 TON AT $ 300.0/TON         =     76200. DOLLARS

NET INVESTMENT COST                             =    123800. DOLLARS

SIMPLE PAYBACK PERIOD                           =      5.37 YEARS

     ******************** INPUT DATA ********************

     SITE:  LOS ANGLES AIRPORT

     OUTSIDE AIR QUANTITY                  =  150000 CFM
     ROOM DESIGN DRY BULB                  =   75.00 DEGREE
     ROOM DESIGN WET BULB                  =   62.50 DEGREE
     COOLING COIL LEAVING DRY BULB         =   55.00 DEGREE
     COOLING COIL LEAVING WET BULB         =   54.50 DEGREE
     WATER COST                            =    1.10 $/1000GAL
     KW PER REFRIGERATION TON              =    1.40 KW/TON
     ELECTRICITY COST                      =     .07 $/KWH
     AIR-TO-AIR HEAT EXCHANGER SP          =     .68 INCH
     FAN EFFICIENCY                        =   80.00%
     SCAVENGER FANS AND PUMP               =   31.60 HP
     SCAVENGER SHUTOFF DRY BULB (OSA)      =   55.00 DEGREE

     ****************************************************

     COOLING IS CONSIDERED SENSIBLE ONLY WHEN HUMIDITY RATIO
     OF THE OUTSIDE AIR IS LOWER THAN THAT OF THE ROOM DESIGN.
     NO HUMIDIFICATION IS CONTEMPLATED.
```

tral chilled water system, the cost avoidance figure is reduced from $350 to $300 per ton.

Indirect evaporative coolers perform better in hot and dry climates. Table 6 shows the results if the laboratory building is moved from the coastal region to San Bernardino, which is a hotter and drier region 60 miles inland. The payback period reduces to 3.81 yr, and the concept can certainly be justified for its energy conservation. If a rooftop unit is used in this case, the payback period is further reduced to 2.28 yr.

Electricity cost plays an important role in justifying the use of an indirect evaporative cooler. To illustrate this point, the energy and water consumption results were extracted from the computer calculation results using a simple spreadsheet program. Table 7 shows the cost impact for the four cases with electricity cost varying from $0.070 to $0.110 per KWH.

Table 6. Indirect Evaporative Coolers Perform Better in Hot and Dry Climates, Shown When This Laboratory Building Is Moved 60 Miles Inland

```
ENERGY CONSUMPTION FOR CONVENTIONAL SYSTEM      =    1170797. KW-HR.
ENFRGY CONSUMPTION WITH PRECOOLING SYSTEM       =     754400. KW-HR.
ENERGY CONSUMPTION SAVING FOR PRECOOLING        =     416397. KW-HR.

ANNUAL ENERGY COST SAVING FOR PRECOOLING        =      29148. DOLLARS

COST OF THE PRECOOLER                           =     200000. DOLLARS
COST SAVING OF 254.0 TON AT $ 350.0/TON         =      88900. DOLLARS

NET INVESTMENT COST                             =     111100. DOLLARS

SIMPLE PAYBACK PERIOD                           =       3.81 YEARS


     ******************** INPUT DATA ********************

     SITE:  SAN BERNADINO

     OUTSIDE AIR QUANTITY                   =  150000 CFM
     ROOM DESIGN DRY BULB                   =   75.00 DEGREE
     ROOM DESIGN WET BULB                   =   62.50 DEGREE
     COOLING COIL LEAVING DRY BULB          =   55.00 DEGREE
     COOLING COIL LEAVING WET BULB          =   54.50 DEGREE
     KW PER REFRIGERATION TON               =     .90 KW/TON
     ELECTRICITY COST                       =     .07 $/KWH
     AIR-TO-AIR HEAT EXCHANGER SP           =     .68 INCH
     FAN EFFICIENCY                         =   80.00%
     SCAVENGER FANS AND PUMP                =   31.60 HP
     SCAVENGER SHUTOFF DRY BULB (OSA)       =   55.00 DEGREE

     ****************************************************


     COOLING IS CONSIDERED SENSIBLE ONLY WHEN HUMIDITY RATIO
     OF THE OUTSIDE AIR IS LOWER THAN THAT OF THE ROOM DESIGN.
     NO HUMIDIFICATION IS CONTEMPLATED.
```

There are a variety of indirect evaporative coolers manufactured today. While the principal theory is the same, the design of each cooler can be different. Consequently, the thermal efficiency of the cooler, which largely depends on the cooler design, can vary significantly. Table 8 shows another cooler, by Manufacturer B, with less thermal efficiency. Even though Cooler B is less expensive than the previous Cooler A, the payback period is significantly longer. Table 9 shows the same electricity cost impact for Cooler B.

Note that when the indirect evaporative cooler is used under the most adverse conditions (i.e., mild weather, high plant *COP*, and low electricity cost), the 31.92-yr payback period for Cooler B is overwhelmingly longer than the 13.98 yr for Cooler A. However, at the other extreme, where all conditions are in favor of using an indirect

Table 7. Spreadsheet for the Four Cases with Electricity Costs Varying from $0.070 to $0.110 per kilowatthour

	LOS ANGELES WITH CONVENTIONAL CHILLER PLANT	LOS ANGELES WITH AIR COOLED ROOFTOP UNIT	SAN BERNARDINO WITH CONVENTIONAL CHILLER PLANT	SAN BERNARDINO WITH AIR COOLED ROOFTOP UNIT
ENERGY SAVING, KWH	113510	342415	416397	798124
WATER UASGE, GALLON	NA	847741	NA	1403457
EQUIPMENT COST, $	200000	200000	200000	200000
COST OF WATER, $/1000 GAL	NA	1.1	NA	1.1
TONNAGE SAVING COST, $	88900	76200	88900	76200
$/KWH	PAYBACK YRS.	PAYBACK YRS.	PAYBACK YRS.	PAYBACK YRS.
0.070	13.98	5.37	3.81	2.28
0.075	13.05	5.00	3.56	2.12
0.080	12.23	4.68	3.34	1.99
0.085	11.51	4.39	3.14	1.87
0.090	10.88	4.14	2.96	1.76
0.095	10.30	3.92	2.81	1.67
0.100	9.79	3.72	2.67	1.58
0.105	9.32	3.54	2.54	1.50
0.110	8.90	3.37	2.43	1.44

Table 8. Indirect Evaporative Cooler Design of Less Thermal Efficiency

```
ENERGY CONSUMPTION FOR CONVENTIONAL SYSTEM     =   1051313. KW-HR.
ENERGY CONSUMPTION WITH PRECOOLING SYSTEM      =   1012777. KW-HR.
ENERGY CONSUMPTION SAVING FOR PRECOOLING       =     38536. KW-HR.

ANNUAL ENERGY COST SAVING FOR PRECOOLING       =      2698. DOLLARS

COST OF THE PRECOOLER                          =    175000. DOLLARS
COST SAVING OF 254.0 TON AT $ 350.0/TON        =     88900. DOLLARS

NET INVESTMENT COST                            =     86100. DOLLARS

SIMPLE PAYBACK PERIOD                          =     31.92 YEARS

      ******************** INPUT DATA ********************

      SITE:  LOS ANGLES AIRPORT

      OUTSIDE AIR QUANTITY              =  150000 CFM
      ROOM DESIGN DRY BULB              =   75.00 DEGREE
      ROOM DESIGN WET BULB              =   62.50 DEGREE
      COOLING COIL LEAVING DRY BULB     =   55.00 DEGREE
      COOLING COIL LEAVING WET BULB     =   54.50 DEGREE
      KW PER REFRIGERATION TON          =     .90 KW/TON
      ELECTRICITY COST                  =     .07 $/KWH
      AIR-TO-AIR HEAT EXCHANGER SP      =     .80 INCH
      FAN EFFICIENCY                    =   80.00%
      SCAVENGER FANS AND PUMP           =   26.70 HP
      SCAVENGER SHUTOFF DRY BULB (OSA)  =   55.00 DEGREE

      ****************************************************

      COOLING IS CONSIDERED SENSIBLE ONLY WHEN HUMIDITY RATIO
      OF THE OUTSIDE AIR IS LOWER THAN THAT OF THE ROOM DESIGN.
      NO HUMIDIFICATION IS CONTEMPLATED.
```

Table 9. Electricity Cost Impact for Indirect Evaporative Cooler Shown in Table 8

	LOS ANGELES WITH CONVENTIONAL CHILLER PLANT	LOS ANGELES WITH AIR COOLED ROOFTOP UNIT	SAN BERNARDINO WITH CONVENTIONAL CHILLER PLANT	SAN BERNARDINO WITH AIR COOLED ROOFTOP UNIT
ENERGY SAVING, KWH	38536	224175	283426	592029
WATER UASGE, GALLON	NA	847741	NA	1403457
EQUIPMENT COST, $	175000	175000	175000	175000
COST OF WATER, $/1000 GAL	NA	1.1	NA	1.1
TONNAGE SAVING COST, $	88900	76200	88900	76200
$/KWH	PAYBACK YRS.	PAYBACK YRS.	PAYBACK YRS.	PAYBACK YRS.
0.070	31.92	6.69	4.33	2.47
0.075	29.79	6.22	4.04	2.30
0.080	27.93	5.81	3.79	2.15
0.085	26.29	5.45	3.57	2.02
0.090	24.83	5.13	3.37	1.91
0.095	23.52	4.85	3.19	1.80
0.100	22.34	4.60	3.03	1.71
0.105	21.28	4.37	2.89	1.63
0.110	20.31	4.16	2.76	1.55

evaporative cooler, the differences in payback periods between the two coolers become relatively insignificant.

Conclusion

The use of an indirect evaporative cooler will conserve energy. The justification of using the cooler, however, must be carefully analyzed. Weather conditions, overall *COP* of the refrigeration system, electricity cost, thermal efficiency of the cooler, and power usage of the cooler accessories all play important roles in determining the effectiveness of the cooler application.

With the aid of small computer programs that can be written by the engineer for the specific application, these evaluations can be carried out in great detail. Once the program is written, running it with varying input combinations or playing the "what if" game takes minimum engineering effort and practically no time.

Appendix **E**

Decision Making in Energy Retrofit Design*

The task of evaluating energy-conservation measures is a complex one. In this retrofit example, the HVAC engineer must analyze numerous alternatives to maximize payback and minimize costs.

HVAC engineers have been retrofitting existing air-conditioning systems for energy conservation ever since the Arab oil embargo of 1973–74. Energy conservation has been so popular in the last decade that the term "energy engineer" emerged to describe a unique group of engineers who specialize in evaluating energy-conservation measures for a given project, new or retrofit.

By definition, energy conservation is an integral part of the HVAC engineer's job. An HVAC engineer, whether he is also titled "energy engineer" or not, must be capable of evaluating the pros and cons of various energy-conservation measures that can or cannot be applied to the building he is commissioned to design.

The task of evaluating energy-conservation measures is multifaceted and complex. This is particularly true for retrofit projects. It is easy to make recommendations to save energy on items where very little first cost is involved. However, where energy-conservation measures

*Reprinted with permission from *Heating/Piping/Air Conditioning,* September 1986.

involve large sums of initial cost, the relationship between the energy cost savings and the initial expenditure must be taken into consideration. Often, it can be proved unwise to spend large amounts of retrofit monies to save a relatively small amount of energy.

The purpose of this article is to discuss the issues involved in the decision-making process when evaluating various energy-conservation measures. A retrofit project is used as an illustrative example.

HVAC System Design

The use of refrigeration principles to cool air for space comfort has been with us for a long time. In the beginning, the reheat method, i.e., heating a constant temperature cold airstream to higher temperatures, was the primary method used to modify the supply air temperature to satisfy individual space needs. During the 1950s and 60s, the constant volume double duct (CVDD) concept was introduced and became a very popular air-conditioning system for high-rise office buildings. The system offers practically trouble-free air conditioning. It is easy to maintain and easy to modify for tenant improvements, and in the meantime, it is not as energy wasteful as the constant volume reheat (CVRH) system. Variable air volume (VAV) systems were not fully developed at that time.

During the late 1960s and 1970s, as VAV systems developed into practical air-conditioning systems, engineers became aware that CVDD systems, although better than CVRH systems, also waste energy in their normal operation. This is particularly true during the heating season when both heating and cooling are activated.

As energy conservation gained popularity, the use of VAV systems became fashionable as if "VAV" were synonymous with "energy conservation." It was believed that a building equipped with VAV systems must be energy conserving. In an energy retrofit project, an engineer could not go wrong by changing a constant volume system to a variable volume system to conserve energy. This article points out that the second part of this statement may not always be true.

CVDD System Energy Wastage

A conventional CVDD system wastes thermal and transportation energy. The system features a cooled air stream and a heated air stream. The space temperature is controlled by supplying cooled or heated air only when a space is subject to design cooling or heating loads, respectively. Under partial load conditions, however, the cooled and heated air streams are proportionally mixed to meet the space load. This mixing action is a thermal energy waste. Generally, energy is not wasted during the cooling season, since the heating coil in the hot duct can be deactivated, and the mixed return air and outdoor air flows through the hot duct.

As the name implies, the constant volume (CV) system is designed to supply a constant amount of air through the system at all times. Space load conditions are met by varying the supply air temperature delivered to each conditioned space. In this type of system, the energy required for the fan motor is constant. Fan motor energy can be drastically reduced if the supply air volume can be reduced under a light-load condition. Such is the case with a VAV system. Compared to a VAV system, a constant volume system uses more, i.e., wastes, transporation energy.

To retrofit a CVDD system for energy conservation, one generally aims at minimizing the thermal energy waste and/or reducing the transportation energy usage.

Retrofit Project Alternatives

The retrofit example is a 328,000 sq ft, 15-story office building with 14 typical floors located in Los Angeles. The building is an all-electric facility. Each floor is served by a separate CVDD air handling unit supplied with chilled water from a central cooling plant located in the penthouse. Heating is provided by a multistage electric heater in the hot plenum. There are approximately 40 CVDD mixing terminals serving each floor. The retrofit study was performed for a typical floor only. Fig. 1 shows a schematic diagram of the existing conventional CVDD system serving each floor of the building. Three retrofit alternatives were considered.

Alternative 1—Constant volume double duct with subzone heaters. The first energy-conservation retrofit alternative is to convert the existing system to a constant volume system with subzone heaters (CVSZH), as shown in Fig. 2. In this alternative, the heater in the hot plenum will be disconnected and the "hot" duct becomes a "bypass" duct at all times. "Subzone heaters" will be installed in the supply duct downstream from the mixing terminals serving the exterior spaces.

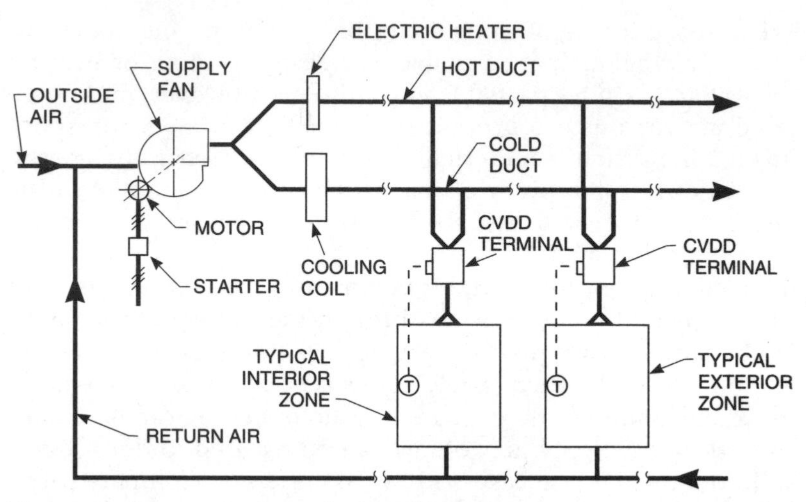

Figure 1. Constant volume double-duct system.

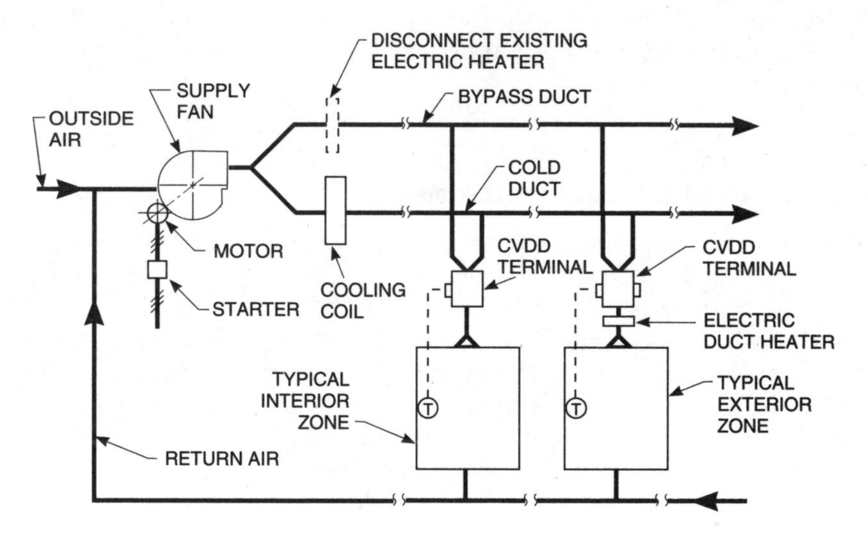

Figure 2. Constant volume double-duct with subzone heaters.

With this modification, heating will not be activated until the cold air damper in the mixing terminal is completely closed. Thus thermal energy wastage due to the mixing of hot and cold air is completely eliminated. Since there will be no heating required in the interior area at any time, no heaters need to be installed in the ducts serving the interior zones. The temperature control for the interior spaces can be accomplished by mixing varying amounts of cold air and bypass air.

No modification is required at the fan for this alternative since the system still delivers a constant volume of air to all conditioned spaces. Therefore, while this alternative eliminates thermal energy wastage, it does not reduce transportation energy waste.

The modifications required for this alternative are primarily the installation of electric duct heaters downstream from the mixing terminals, wiring the heaters to the electrical panel, and modifying the controls at the terminal to activate the heaters. The total installation cost for this modification is estimated at $17,500 for a typical floor.

Alternative 2—Variable volume double-duct system. The second alternative is convert the existing system to a variable volume double-duct system (VVDD) as shown in Fig. 3. In this alternative, the heater in the hot plenum will remain. All of the constant volume mixing terminals will be changed to the variable volume type. As the space cooling load reduces, the mixing terminal reduces the cold air volume only without opening the hot air port. When the cold air quantity decreases to a preset minimum, further load reduction opens the hot air port and starts the mixing action. Heating will be provided by the heater in the hot plenum. With this modification, the amount of thermal wastage is reduced, since the amount of air involved in the mixing action is reduced.

Cooling-only VAV terminals will be used for interior zones, since there will be no heating required in the interior area at any time. As long as the minimum ventilation rate does not have to be maintained at all times, these cooling-only VAV terminals will reduce the supply air volume to zero under a no-load condition.

A variable-speed drive and its associated controls will be installed for the supply fan to decrease the amount of supply air as the air quantity required at the terminals reduces. While there are other methods to reduce the supply air volume, the variable-speed drive is the easiest to install in a retrofit project and offers the greatest fan energy reduction. With this modification, the thermal energy wastage is minimized, and the transportation energy from the fan motor is drastically reduced because of the VAV operation.

The modifications required for this alternative include the installation of the variable-speed drive at the air handling unit and the changing of all CV terminals to VAV terminals. An additional control is also needed to control the variable speed drive in response to the system needs. The total installation cost for this alternative is estimated at $36,200 for a typical floor.

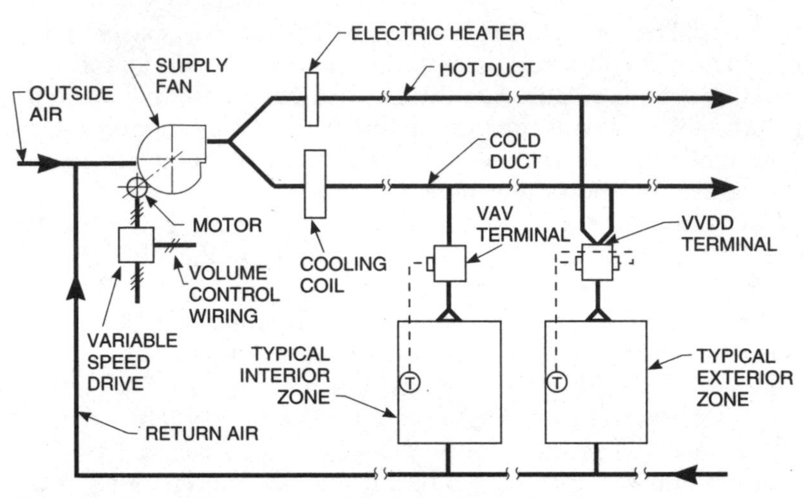

Figure 3. Variable volume double-duct system.

Alternative 3—Variable volume double duct with subzone heaters. The third alternative is to convert the existing system to a variable volume double duct with subzone heaters (VAVSZH), as shown in Fig. 4. This alternative combines the energy conservation features of the first two alternatives. Consequently, it will also require all of the modifications for the first two alternatives.

In this alternative, as the space cooling load reduces, the mixing terminal reduces the cold air volume only without opening the bypass air port. When the cold air quantity decreases to a preset minimum, further load reduction opens the bypass air port and starts to mix the cold air with the bypass air. Heating will not be activated until the cold air damper in the mixing terminal is completely closed. With this modification, not only is the amount of thermal wastage eliminated, but the transportation energy is also reduced, since the amount of supply air is reduced as the load decreases.

As expected, this alternative will save the greatest amount of energy, and equally as expected, this alternative will require the highest initial cost. The installation cost for this alternative is estimated at $51,700 for a typical floor.

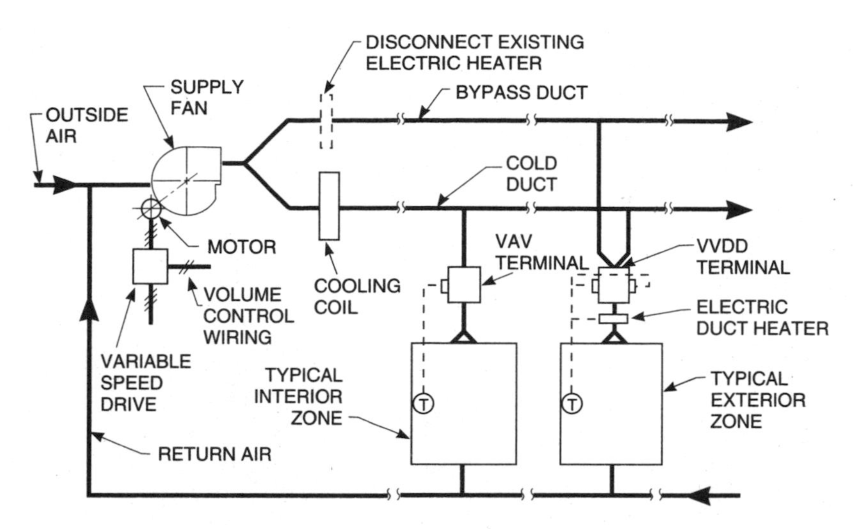

Figure 4. Variable volume double-duct with subzone heaters.

Energy Cost Estimates

To evaluate the effectiveness of these alternatives, energy consumptions for each alternative must be calculated. An energy simulation computer program was developed to estimate the energy consumptions for three alternatives.

For a given HVAC system, the energy simulation program calculates hourly heating and cooling loads and the motor horsepower required to cope with these loads, taking into account the external and internal load variations as well as the thermal and transportation efficiencies of the system. These hourly loads are totalized monthly based on the number of operating hours per day and the number of operating days in each month. The annual energy consumption is the sum of accumulated monthly consumptions.

Table 1 and Fig. 5 summarize and illustrate for a typical floor the simulated annual energy consumptions for heating, cooling, and the fan motor for the three alternatives together with the CVDD base case. As expected, the results show that among the three alternatives considered, the VAVSZH system will achieve the greatest amount of energy savings.

Table 1. Simulated Energy Consumption for a Typical Floor

System	Heating, kWh	Cooling, kWh	Fan motor, kWh	Total, kWh	Total kWh 14 floors
CVDD	50,727	99,222	49,694	199,643	2,795,000
CVSZH	11,465	92,749	49,694	153,908	2,154,700
VAVDD	24,913	90,046	28,916	143,875	2,014,250
VAVSZH	14,547	88,287	28,916	131,750	1,844,500

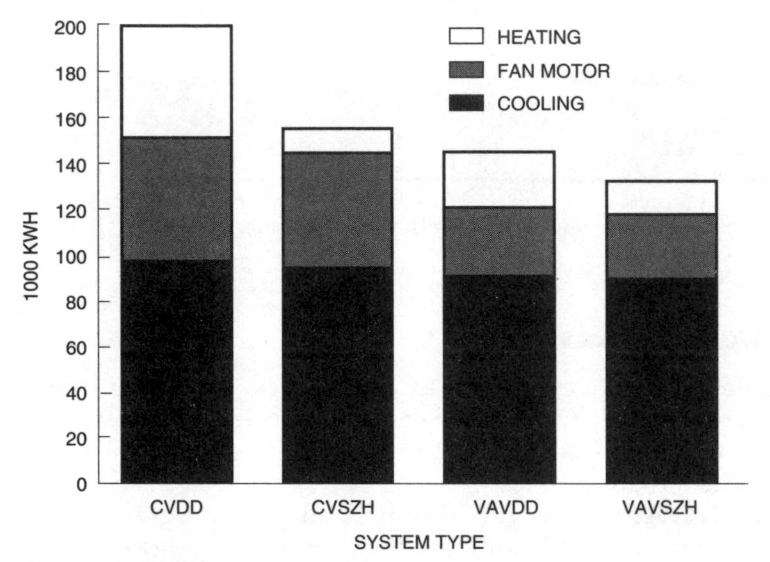

Figure 5. Energy consumption comparison.

Overall Consideration

As stated at the beginning, energy-conservation considerations cannot be stopped at saving energy alone. While it is easy to decide on energy retrofit items where very little initial costs are involved, it can be foolhardy to spend large sums of money to save very small amounts of energy. Between these two extremes, there lies a wide spectrum of possibilities requiring careful consideration.

Table 2 tabulates the initial costs, annual energy, and energy cost savings and simple payback periods for the three alternatives applied to the 14 typical floors. Energy cost is calculated using an average electricity cost of $0.08 per KWH. It can be seen that while the VAVSZH

Table 2. Comparison of Energy-Conservation Alternatives

System	First Cost	Energy savings kWh	Energy savings Costs	Simple payback period, years
CVSZH	$245,000	640,300	$51,220	4.8
VAVDD	$506,800	780,750	$62,460	8.1
VAVSZH	$723,800	950,500	$76,040	9.5

alternative saves approximately 50 percent more annual energy cost than the CVSZH alternative ($76,020 versus $51,240), the initial retrofit cost of the VAVSZH alternative is nearly 300 percent higher than that of the CVSZH alternative ($723,800 versus $245,000). To put it another way, while it is relatively easy to justify the expenditure for $245,000 to save $51,240 annual energy cost, it may be difficult to justify spending an additional $478,800 to save the extra $24,780 annually.

It is interesting to note that the only difference between the CVSZH and VAVSZH alternatives is the addition of VAV operation. While it is true that a VAV system conserves energy, it is probably unwise and not justifiable in this case to retrofit the CVDD system to a VAVSZH system.

Payback Period

A crude indicator to judge the applicability of an energy retrofit alternative is the use of a simple payback period. In this case, the payback periods in Table 2 are calculated simply by dividing the initial costs by the cost of annual energy savings. The discount rate and the escalation of energy costs are not taken into consideration. Many institutions and government agencies use some form of simple payback period as the preliminary indicator to judge the desirability of the retrofit alternatives.

Most of the simple payback calculations ignore factors such as maintenance costs and the salvage value of the equipment. These factors, when taken into consideration, can either confirm or sometimes alter the ranking order of the alternatives. In the example cited in this appendix, the periodic maintenance cost of the variable speed drive, which is not needed in the constant volume alternative, will further erode the energy cost savings of the VAVSZH system.

Conclusion

Energy conservation, particularly retrofitting for energy conservation, is not just to save Btus. With a few exceptions, every conservation measure has a price tag. Buying energy conservation is not like buying a diamond ring; practical decisions have to be made by evaluating alternatives based on the value of the investments.

Many energy conservation measures are accumulative. Most of the time, it is always easier to save a large amount of energy for a small amount of investment at the beginning. As refined measures are added, the return on investment diminishes. There are cases where trying to save the last Btu becomes a poor investment.

Index

*Note: The *f*. after a page number refers to a figure.